Fixed Stars
and
Judicial Astrology

George Noonan

Copyright 1990 by American Federation of Astrologers, Inc.
All rights reserved.

No part of this book may be reproduced or transmitted in any form or by any means, electronic or mechanical, including photocopying or recording, or by any information storage and retrieval system, without written permission from the author and publisher. Requests and inquiries may be mailed to: American Federation of Astrologers, Inc., 6535 S. Rural Road, Tempe, AZ 85283.

ISBN-10: 0-86690-376-3
ISBN-13: 978-0-86690-376-9

First Printing: 1990
Current Printing: 2009

Cover Design: Jack Cipolla

Published by:
American Federation of Astrologers, Inc.
6535 S. Rural Road
Tempe, AZ 85285-2040

Printed in the United States of America

Contents

Introduction	v
The Nature of Fixed Stars	1
The Northern Constellations	7
The Zodiacal Constellations	33
The Southern Constellations	57
Genethiacal Applications of the Fixed Stars	79
Judicial Astrology	85
Appendix A, The Mathematics of Astrology	111
Appendix B, Rulership of a Point in a Chart	127
Appendix C, The Stars You Are Born Under	131

Introduction

Perhaps the least understood and most misused elements in astrology today are the fixed stars. Current practice is to consider the stars as no more than mini-manifestations of the planets whose nature they are presumed to hold. The location of the stars projected onto the ecliptic are then considered valid points for aspectual relationships with the actual planets. No special significance is attributed to the non-zodiacal constellations, nor is a star's location within a constellation given any weight whatsoever. Finally, modern astrology places no distinction on the use of the stars and constellations in genethliacal astrology and the other branches of the ancient art.

The only correspondence of modern practice with classical astrology is that the stars do indeed have planetary nature. In almost all other respects modern astrology has no relationship at all with its classical counterpart. To begin with, the classicists did not project the position of the stars onto the ecliptic, but rather maintained their locations in the sky as observed from the Earth. Nor were the stars considered merely a minor manifestation of the planets whose nature they possessed. The stars in classical astrology are an extremely important element in judicial (mundane) astrology, even surpassing the planets in many regards. And their use in genethliacal and judicial astrology is considerably different. Finally, the constellations and the stars's positions within them is explicitly considered in all delineations. For example, in some applications the constellation of Orion is considered more important than any of the signs of the zodiac!

In this book the natures of the fixed stars and constellations as known by the classicists will be discussed in detail. In this regard ancient star catalogs will be used and the ancient stellar designations, which were descriptive, will be put into one-to-one correspondence with the modern catalog designations. The ancient constellations will be described, and the differences with modern constellations will be noted. Methods of delineating the fixed stars will be indicated, with a special emphasis on the applications to judicial astrology.

As mentioned above, the ancient stellar designations have been put in one-to-one correspondence with the modern ones. As for example, "The star in the left knee of the Man Kneeling" is given the designation of theta-Herculis in modern catalogs. For the brighter named stars this transformation presents no problem. But many of the 1,022 stars in the ancient catalog are of magnitude four and five, and the specific star meant is ambiguous. Every effort has been made to assure that these stars are correctly placed (e.g. the star in the right flank of the Man Kneeling has been assumed to be the magnitude 4.6 star upsilon-Herculis). It is recognized, however, that for some of these dimmer stars there may be errors in their modern designations. Also, in some instances their location is so ambiguous that all anyone can do is guess. These latter case are less than five percent of the total. It is expected that the reader will discover most of these anomalies (if any), which can be corrected in an addenda.

This volume is a sequel to its companion *Classical Scientific Astrology*. It begins where the other left off, but it is not absolutely necessary for the reader to be familiar with the first volume. With few exceptions, any required information from *Classical Scientific Astrology* necessary for the understanding of this book will be included herein. It is recommended that the review questions at the end of each chapter be answered. They are designed to add to the basic text.

The first portion of Appendix A is an edited repeat of a little booklet published previously, *Spherical Astronomy of Astrologers*. It is replicated here to ensure completeness and as a background for the second half of the Appendix. The latter portion of Appendix A is completely new material that will be of value to those who wish to put into practice the theories of classical astrology. It is intended that this book and its companion mentioned above be considered as a single volume incorporating the major tenets of classical astrology.

George Noonan
Sylmar, California

Chapter I

The Nature of the Fixed Stars

The precession of the equinoxes has slowly moved the stars backward in the sky so that today the beginning of spring is marked not by the first part of the constellation of Aries, but by those stars near the beginning of the constellation of Pisces. This lack of correspondence between the signs and the constellations forms the basis for both the criticism of astrology by its modern opponents and the formation of the modern sidereal school of astrology. As is indicated in *Classical Scientific Astrology*, classical astrologers believed that the signs were related to the seasons and so disregarded precessional effects when developing the natures of the signs. The modern criticism of astrology based on the precession of the equinoxes is therefore not relevant.

Sidereal astrologers hold to twelve signs, each of thirty degrees in extent, and in one-to-one correspondence with the constellations of the same name. Currently they place the vernal equinox at six degrees in the sign (and constellation) of Pisces, but in other matters they proceed generally as do the tropical astrologers. (There are minor differences such as the interpretation of the aspects and their rejection of planetary rulership of the signs—domiciles of the signs in classical terminology—but a complete discussion would take us too far afield.) Not only do the modern sidereal astrologers completely neglect much of the astrological theory of the classicists, but err in claiming that the zodiacal constellations are each thirty degrees. In fact, the ancient constellations range in size from 18°30' for the constellation of The Crab (Cancer) to 46°20N for the constellation of The Virgin (Virgo). The sum of the sizes of all the zodiacal constellations is less than 360 degrees, indicating the amount of irregularity and overlap involved. This information comes from ancient star catalogs (Ptolemy, "*Almagest*, Great Books of the Western World," Vol. 16, *Encyclopedia Brittanica*, Chicago, 1952), but many siderealists assert that the ancient books are all wrong (private correspondence with the founders of a society of sidereal astrologers). They rely instead on modern books written by those who lack any knowledge of the history of science or of classical astrology. Such has astrology deteriorated from its proud heritage. Actually, as will be seen later, classical astrology through its use of the constellations is also a sidereal astrology. In classical astrology the precession of the constellations is recognized. The signs, however, remain constant with their beginning at the vernal equinox.

The constellations (as opposed to the sign) and the fixed stars play a prominent role in classical astrology, especially in judicial astrology (see Chapter VI). Ptolemy devoted two large chapters to the natures of the fixed stars and to the effects of the constellations and makes other references to these important elements throughout *Tetrabiblos*. The astrological natures of the fixed stars will be described in the following three chapters, and their major applications in Chapters V and VI. However, it is first necessary to digress a bit to explain the manner in which these elements become important in a chart.

As explained in *Classical Scientific Astrology*, the planets exert their influence through aspectual relationships with each other, with the signs, and with important points in the chart such as the horoscope Ascendant and Midheaven. These planetary aspects are not in general applicable to the fixed stars. Remember that in classical astrology the positions of the stars is not translated to the ecliptic. They keep their position in the heavens as seen from the Earth. As a result most of the stars are so far removed from the ecliptic that a classicist would say "their rays cannot co-mingle." And, second, the classical astrologers applying observational astronomy would naturally utilize those relationships most visually apparent: the rising, setting, and culmination of the stars. Those stars that are rising, setting, and culminating at the time and place for which a nativity is cast are those stars that the native is born under. These are extremely important as will be seen in Chapter V. The Appendix shows how to determine these stars and also indicates the manner in which they should be delineated.

But there are also stellar aspects, and Ptolemy explicitly refers (in viii:4 of the *Almagest*) to aspects of the fixed stars with the Sun when indicating the applications of the fixed stars and constellations to the correlative effects of eclipses (in ii:7 of *Tetrabiblos*). A star that is rising in the morning with the Sun was said by Ptolemy to be *matutine subsolar*. That star rising with the Sun on the day and place that the chart is cast for is therefore in a *matutine subsolar* aspect with the Sun. When the Sun is rising a star may also be culminating or setting. These aspects are called *matutine culmination* and *matutine setting*. Hence a star that is culminating or setting at the instant the Sun is rising on the day for which a chart is cast is said to be in these aspects with the Sun.

Now place the Sun on the Midheaven on the day and place for which the chart is cast. Stars that are rising, culminating, and setting as the Sun is culminating are said to be in *meridianal subsolar*, *meridianal culmination*, or in a *meridianal setting* aspect with the Sun. Finally, stars that are rising, culminating, and setting as the Sun is setting are in a *vespertine rising*, *vespertine culmination*, and *vespertine setting* aspect with the Sun.

Stellar aspects with planets other than the Sun are restricted to co-rising, co-culmination, and co-setting of the star and planet. An exception is occultation of a star by the Moon or planets. But this latter aspect, which has the force of a planetary conjunction, can only occur with those stars close to the ecliptic. There is a natural extension of the six stellar aspects with the Sun to the planets, but there is no evidence that they were used in classical times.

The rising and culminating aspects are more benefic (or at least more significant). As a setting star is apt to be connected with death, these aspects are more malefic. Matutine aspect are stronger than meridinal, which are stronger than vespertine. The actual influence of a star or constellation is dependent on its nature in combination with its aspect.

The astrological natures of the stars are akin to those of the planets. They are given by Ptolemy in *Tetrabiblos*, albeit in a most cursory manner. He writes:

> "...the bright stars in the constellation of the Little Bear have a similar quality to that of Saturn, and to a less degree, to that of Venus...(those) bright stars in the constellation of the Dragon, to that of Saturn, Mars, and Jupiter."

Modern astrologers have taken this to mean that since Polaris (α-UMi) is the brightest star of those in Little Bear that it must therefore have the nature of both Saturn and Venus, and likewise Alwaid (ρ-Dra) must have the nature of Saturn, Mars, and Jupiter combined. (In *Fixed Stars and Constellations in Astrology*, Vivian Robson also mistakes Rastaban for Alwaid and makes many other errors.) This is all very logical. Right? Wrong!

To begin with, Ptolemy did not even consider Polaris the brightest star in Ursa Minor, listing it as a dimmer than both Kochab (β-UMi) and Pherkad (γ-UMi). The star closest to the north pole at the time Ptolemy wrote was Kochab, and Polaris was not even mentioned in the literature until after 500 A.D. Furthermore, reading classical astrology in context indicates that the classicists did not consider the stars in general to have the natures of a multitude of planets individually. What Ptolemy meant is that in the constellation of Little Bear there are stars with the nature of Saturn and that there are other stars of the nature of Venus. Of course, these same remarks are true of the constellation of The Dragon as well. But Ptolemy mentions very few stars by name, and some of these do have multiple natures. Hence the confusion that permeates modern astrology in this matter.

The natures of the stars were assigned by the classicists on the basis of their color: red to Mars and Saturn, white to Venus, blue to Jupiter and the Moon, and yellow to Mercury and the Sun. It is not surprising, therefore, to find a strong correlation of the ancient natures of the fixed stars with their modern stellar spectra. In this regard the stars can be classified according to their spectra in a very natural way from the most benefic to the most malefic.

Class O stars are those whose spectra is very strong in the ultraviolet lines of nitrogen and life-giving oxygen. Stars in this class would have the nature of the most benefic planet of all—the Moon. Class B stars are strong in helium, but also with nitrogen and oxygen. They are astrologically as Jupiter. Class A stars are those in which the element hydrogen is prominent. They are as Venus. Mercury is neutral as to a benefic or malefic nature, and stars with a spectral class F can also be classified as neutral in that this class contains minor metals such as calcium and is a bridge to the more metallic stars. So stars in spectral class F are as Mercury. Our Sun is a spectral class G star. Surprisingly (or is it so?), in the natural order in which we have been listing the stars, their spectra, and their astrological natures, stars in spectral class G are astrologically as the Sun! Metallic lines, especially iron, dominate the spectra of class K stars. Class K stars are therefore of the nature of Mars. Finally, those stars in spectral class M, with its lines of harsh metals such as titanium oxide, have the astrological nature of the highly malefic Saturn. To summarize:

Stellar Spectra	Planetary Nature
O	Moon
B	Jupiter
A	Venus
F	Mercury
G	Sun
K	Mars
M	Saturn

It will be seen that this scheme follows that of Ptolemy quite closely. In the few instances in which it results in a star being given a nature at variance with that of the classicists, the nature as predicted by the spectra is most generally confirmed by the lore associated with the star. Of course Ptolemy never gave the natures of Uranus, Neptune, or Pluto to any of the stars. These planets were unknown in his time. Perhaps future research will rectify this situation and find stars with the nature of one or more of the trans-Saturnian planets.

Ptolemy lists forty-eight constellations and 1,022 stars in his catalog. Few of these stars have ever been given names, and fewer still have a history of astrological relevance. In what follows, each other constellations will be discussed in turn. Ptolemy's designation for each of the stars will be given along with its corresponding modern designation. In addition, spectral type, apparent magnitude, right ascension, and declination will be given for the named star that are most likely to be of astrological importance. The Appendix lists formulae for computing stellar aspects, and subsequent chapters will deal with specific applications of the fixed stars and constellations. Please note that stellar positions have changed in the past two millennia so that the figures described by Ptolemy will be distorted in the sky today. Also, the ancient constellations have been changed in modern catalogs. These changes are noted where they occur.

Tables 1 and 2 present the format in which the star will be discussed in the following chapters. First, all the stars in the constellations as listed by Ptolemy in his catalog in the *Almagest* are listed together with their modern designations. For the constellation of the Little Bear (modern Ursa Minor) Ptolemy lists eight stars, seven of which are in the constellation proper, and one of which is unfigured. The stars in the constellation proper are delineated, taking into account the nature of the constellation of which they are a member. The unfigured stars do not partake of the

Table 1
Constellation of the Little Bear

The Stars According to Ptolemy	Modern Designation
The star at the tip of the tail	α-UMi
The next one in the tail	δ-UMi
The next one, before the beginning of the tail	ε-UMi
The southern one on the western side of the rectangle	η-UMi
The northern one on the same side	ζ-UMi
The southern one of those on the eastern side	γ-UMi
The northern one on the same side	ρ-UMi
The Unfigured Star	
The more southern star in a straight line with those on the eastern side	5-UMi

Table 2
Named Stars of the Little Bear

Designation	Name	Type	Apparent Magnitude	Position RA	Decl.
α-UMi	Polaris	F8	2.1	02 07	89°08'
ρ-UMi	Kochab	K5	2.2	14 51	74°17'
δ-UMi	Pherkard (Yildun)	A0	4.4	17 43	86°36'
γ-UMi	Pherkad	A2	3.1	15 21	71°56'

constellational nature quite as much. For example, The Little Bear has an evil reputation (see Chapter II), so even though Pherkard and Pherkad have the nature of Venus, it would be the harsher portion of the Venusian nature that would be stressed with these stars.

For each constellation there is also a table of the named stars (Table 2). This table contains the modern stellar designation, the star's name, its location in right ascension and declination, its apparent magnitude, and its spectral type. From the spectral type the astrological nature of the star can be known. Stellar spectra are rated zero through nine. K0 means that a star is pure Mars. B5 means that the star is beginning to shade from Jupiter toward Venus. These subtleties of meaning can be accounted for in a delineation. The star's location in right ascension and declination is required to determine when it will rise, culminate, and set (see Appendix). The lower the apparent magnitude the brighter the star (a magnitude two star is ten times brighter than a magnitude three star). In classical astrology the brighter a star (or a planet) in the sky the more important it becomes. A dim star (or at times even the planet Venus) may be beneficial by nature, but its lack of brightness dims this effect in a chart.

Review Questions

1. Pherkad has a celestial longitude of 21 Aquarius 12 and a celestial latitude of 75N15. What aspects can this star make with a planet at latitude 0°0'? Explain.

2. Pherkard and Pherkad both have the nature of Venus. What would be the difference between them if located on the Midheaven?

3. Describe the nature of the star Polaris.

Chapter II

The Northern Constellations

The Constellation of the Little and Big Bear

These constellations are today called Ursa Minor (UMi) and Ursa Major (UMa) respectively. In ancient times the Big Bear was more extensive than modern Ursa Major, containing the stars of the constellations Canes Ventici (CVn), Lynx (Lyn), and Leo Minor (LMi).

Both the Little and Big Bear are circumpolar and never rise or set in the latitudes of interest to most modern astrologers. Hence the only applicable aspects for the stars of these constellations are the culminations Ptolemy says that the bright stars of the Little Bear are astrologically as Saturn and Venus, and those of the Big Bear as Mars. The unfigured stars beneath the tail of the Big Bear in what is today the constellation of Canes Ventici have the nature of the Moon and Venus.

Astrologically both Bears were said to presage an evil influence. They are particularly injurious as regards to the affairs of nations and kings. However, note that the unfigured stars, especially those in Canes Ventici, can be quite beneficial regarding these matters.

We have already mentioned the major star of the Little Bear. In addition to assigning (erroneously) a dual nature of Saturn and Venus to Polaris, Robson (in *The Fixed Stars and Constellations in Astrology*) also asserts that "the Arabs were of the opinion that the contemplation of Polaris cured ophthalmia. He got this wrong too as it was the star al-Kiblah (or Kochab) for which this power was asserted.

Stars of note in the Big Bear include Mizar which is said to presage the death of a loved one by the ancient astrologers.

Table 3—The Constellation of the Big Bear

The Stars According to Ptolemy	Modern Designation
The star at the tip of the muzzle	o-UMa
The western star of those in the two eyes	π₂-UMa
The eastern one of these	π₁-UMa
The western star of the two in the forehead	ρ-UMa
The eastern one of these	ξ-UMa
The star at the end of the western ear	24-UMa
The western star of the two in the neck	ζ-UMa
The eastern one of these	23-UMa
The northern star of the two in the breast	υ-UMa
The southern one of these	φ-UMa
The star in the left knee	θ-UMa
The northern star at the end of the left forefoot	ι-UMa
The southern one of these	κ-UMa
The star above the right knee	18-UMa
The star below the right knee	15-UMa
Of those in the quadrilateral, the star on the back	α-UMa
Of these, the star on the flank	β-UMa
The star at the beginning of the tail	δ-UMa
The remaining star in the left thigh	γ-UMa
The western star of those at the end of the left hind foot	λ-UMa
The star east of this one	μ-UMa
The star in the left ham	ψ-UMa
The northern star of those at the end of the right hind foot	ν-UMa
The southern star of these	ξ-UMa
The first star of the three in the tail after the beginning	ε-UMa
The middle one of these	τ-UMa
The star at the end of the tail	η-UMa

The Unfigured Stars

The star under the tail far to the south	α-CVn
The dimmer star west of it	β-CVn
The more southern star of those between the forefeet of the Bear and the head of the Lion	α-Lyn
The star north of this	38-Lyn
The star east of the other three dim ones	13-LMi
The star west of this last	11-LMi
The star still farther west than this last	10-LMi
The star between the forefeet and the Twin	31-Lyn

Table 4—Named Stars in The Big Bear

Designation	Name	Type	Apparent Magnitude	Position RA	Decl.
ε-UMa	Aloith	A0	1.7	12h 53m	56° 06'
η-UMa	Alkaid	B3	1.9	13 47	49 27
ξ-UMa	Alula Austrlis	G0	3.9	11 17	31 41
ν-UMa	Alula Borealis	K0	3.7	11 17	33 14
β-UMa	Asterion	G0	4.3	12 33	41 30
α-UMa	Chara (Cor Caroli)	A0	2.9	12 55	38 28
α-UMa	Dubhe	K0	1.9	11 02	61 54
δ-UMa	Megrez (Kaffa)	A2	3.4	12 14	57 11
ρ-UMa	Merak	A0	2.4	11 01	56 30
ζ-UMa	Mizar	A2	2.4	13 23	55 04
o-UMa	Museida	G0	3.5	08 28	60 48
γ-UMa	Phecda (Phad)	A0	2.5	11 52	53 50
ι-UMa	Talitha	A5	3.1	08 57	48 08
μ-UMa	Tania Australis	K5	3.2	10 21	41 38
λ-UMa	Tania Borealis	A2	3.5	10 15	43 04

Alkaid was called "the destroyer of nations" by Islamic astrologers, but it is not known if this was only related to the "favorable" destruction of enemy nations or pertained to nations in general. Alkaid has the nature of Jupiter so the former interpretation is the most likely.

The star Chara has a most benefic reputation as regards the heads of state. It was said to have shown with particular brilliance when Charles II returned to London on May 29, 1660.

Dubhe (α-UMa), when rising, is said to presage those who are kind to wild beasts and who "associate with bulls and lions as if they were people." When setting and aspected by Mars (Mars aspecting the Descendant) the native is said to be in danger of "being killed by wild beasts."

One named star not found in Ptolemy's catalog is the fourth magnitude star Alcor (80-UMa). It is a companion to Mizar and in Ptolemy's time was too close to Mizar to be distinguished with the naked eye. Later on, however, the Arabs used Alcor as a test of good eyesight. Today the star is brighter than formerly, and farther away from its companion, so that it can be readily seen. It is a Class A5 star, and hence of the nature of Venus.

The Constellation of the Dragon

Draco (Dra) was circumpolar about 5000 B.C., and some of its stars have been pole stars in the past. Ptolemy states that the bright stars of this constellation have the nature of Saturn, Mars, and

Table 5—The Constellation of The Dragon

The Stars According to Ptolemy	Modern Designation
The star on the tongue	μ-Dra
The star in the mouth	ν-Dra
The star above the eye	β-Dra
The star in the jaw	ξ-Dra
The star above the head	γ-Dra
The northern star of the three in a straight line in the fist fold of the neck	39-Dra
The southern one of these	46-Dra
The middle star of these	45-Dra
The star east of this one	o-Dra
The southern star of the western side of the square in the next turn	π-Dra
The northern star of the western side	δ-Dra
The northern star of the eastern side	ε-Dra
The southern star of the eastern side	ρ-Dra
The southern star of the triangle in the next turn	σ-Dra
The western star of the remaining two of the triangle	μ-Dra
The eastern star of these	τ-Dra
The eastern star of those in the triangle west of this	ψ-Dra
The southern star of the remaining two of the triangle	φ-Dra
The northern star of the remaining two	χ-Dra
The eastern star of the two small stars west o the triangle	ω-Dra
The western one of these	27-Dra
The more northern of the next three in a straight line	18-Dra
The middle one of the three	19 and 20-Dra
The northern one of these	ζ-Dra
The northern one of the next two to the west	γ-Dra
The southern one of these	θ-Dra
The western star of those to the west in the turn near the tail	ι-Dra
The western star of the two rather distant from the last one	10-Dra
The eastern star of these	α-Dra
The star near them by the tail	κ-Dra
The remaining star at the end of the tail	λ-Dra

Jupiter. The early astrologers called Draco "The Poisonous Dragon" and asserted that a comet within its borders scattered poison over the world. The constellation is especially hurtful to mineral resources, and presages the pollution of rivers and the air. The Moon's Nodes are named after this constellation. The Dragon's Head (North Node) and the Dragon's Tail (South Node) refer to the Moon's undulating course symbolized by the twisting of the Dragon about the North Pole.

Table 6—Named Stars in The Dragon

Designation	Name	Type	Apparent Magnitude	RA		Decl.	
ζ-Dra	Al-Dhi'bah	B5	3.2	17h	09m	65°	45'
δ-Dra	Al-Tais	K0	3.2	19	13	67	37
ρ-Dra	Alwaid	G0	3.0	17	30	52	19
μ-Dra	Arrakis	F6	5.1	17	04	54	34
ι-Dra	Edasich	K0	3.5	19	23	59	03
ψ-Dra	Dziban	F5	4.9	17	43	72	11
λ-Dra	Gianfar	M0	4.1	11	30	69	28
ξ-Dra	Grumium	K3	3.9	17	53	56	53
ν-Dra	Kuma	A8	5.0	17	32	55	12
γ-Dra	Rastaban (Eltanin)	K5	2.4	17	56	51	29
α-Dra	Thuban	A0	3.7	14	04	64	30
ε-Dra	Tyl	K0	4.0	19	48	70	12

The star Thuban (α-Dra) was the pole star 3000 years ago. It was called the "Judge of Heaven" during the reign of the first Saragon king of Akkad, and it can be seen from the bottom of the central passage of the Great Pyramid of Cheops. When rising the star indicates those who are prospectors of gold and silver, or who are ministers of money. If setting and if Mars is conjunct the Descendant or in a harsh aspect with it, it is said to presage the native being burned in his own house or killed by public execution.

(Remember that classical interpretations were generally very descriptive, tending to the extremes. In modern terminology Thuban, when rising, might presage occupations in banking, insurance, or other financial professions. See also the review questions for other indications of the meanings of ancient interpretations.)

The orange star Rastaban was worshipped (as was Thuban) by the ancient Egyptians. It was the pole star about 3500 B.C. The Karnak temples of Rameses and Khons at Thebes were oriented to this star. Gianfer has the nature of the planet Venus, but its location in the Dragon's Tail has earned it the sobriquet of "the poison place." The fourth century astrologer Firmicus Maternus claimed (in *The Mathesis, Ancient Astrology Theory and Practice*) that this star "indicates snake charmers and preparers of medicine from poisons and pigments of herbs."

The only star most modern astrologers recognize in this constellation is Alwaid (erroneously labeled Rastaban by some). It is asserted that this star gives loss of property, violence, and accidents.

Table 7—the Constellation of Cepheus

The Stars According to Ptolemy	Modern Designation
The star in the right foot	κ-Cep
The star in the left foot	γ-Cep
The star under the belt on the right side	β-Cep
The star touching the right shoulder from above	α-Cep
The star touching the right elbow from above	η-Cep
The star touching the same elbow from below	ο-Cep
The star in the chest	ξ-Cep
The star in the left arm	ι-Cep
The southern star of the three in the tiara	ε-Cep
The middle one of the three	ξ-Cep
The northern one of the three	λ-Cep
The Unfigured Stars	
The star west of the tiara	ν-Cep
The star east of the tiara	δ-Cep

Table 8—Named Stars of Cepheus

Deisgnation	Name	Type	Apparent Magnitude	Position RA	Decl.
α-Cep	Aldermin	A5	2.6	21h 18m	62° 29'
β-Cep	Alphirk	B1	3.3	21 28	70 27
γ-Cep	Alrai	K0	3.4	23 38	77 30
μ-Cep	Erakis	M2	4-5	21 43	58 40

The Constellation of Cepheus

Ptolemy asserts that the bright stars of Cepheus (Cep) have the nature of Saturn and Jupiter. In classical times Cepheus was said to presage earthquakes and other events that affect large portions of mankind. The star Alri will be the pole star in 2600 A.D., and Aldermin will have that honor in 7500 A.D. Using our classification by spectra the stars of this constellation have the natures of Venus (Aldermin), Jupiter (Alphirk), Mars (Alrai), and Saturn (Erakis).

When rising in a geneathical chart the star Aldermin (α-Cep) indicates austere natives—those who always assume the countenance of upright characters, and are feared for their severity. If the horoscope (Ascendant) is aspected by either Venus or Mercury while Aldermin is rising the indications are for a career in the theater as a writer of dramas or as an actor. Aldermin was rising in the charts of both Cato (234-1149 B.C.), called "the Elder," Roman statesman, and Tullian (Tullianus Symmachus Valerius, Roman consul, A.D. 330).

Table 9—The Constellation of the Ploughman

The Stars According to Ptolemy	Modern Designation
The star west of the three in the left hand	ι-Boo
The middle and southern one of the three	κ-Boo
The eastern one of the three	θ-Boo
The star in the left elbow	λ-Boo
The star in the left shoulder	γ-Boo
The star in the head	β-Boo
The star in the right shoulder	δ-Boo
The star north of these and in the crook	μ-Boo
The star north of this at the tip of the crook	ν-Boo
The northern of the two below the shoulder in the cudgel	η-Boo
The southern one of these	χ-Boo
The star at the tip of the right hand	45-Boo
The western one of the two in the wrist	46-Boo
The eastern one of these	ψ-Boo
The star at the end of the crook's haft	ω-Boo
The star in the right thigh in the girdle	ε-Boo
the eastern of the two in the girdle	σ-Boo
The western one of these	ρ-Boo
The star in the right heel	ξ-Boo
The northern one of the three in the left shank	η-Boo
The middle one of the three	τ-Boo
The southern one of these	ν-Boo
The Unfigured Star	
The fiery star called Arcturus between the thighs	α-Boo

The Constellation of The Ploughman

Today this constellation is called Bootes (Boo). Its stars traditionally are said to have the nature of Mercury and Saturn. The fiery Arcturus (see below) was given the double nature of Jupiter and Mars by Ptolemy in the *Tetrabiblos*, but the passage is ambiguous and it is likely that the star Nekker (β-Boo) is meant in reference to Jupiter. This constellation is correlative of events concerning agriculture and the effects concerned with this constellation are notably delayed in coming about.

Arcturus (α-Boo) was one of the first stars to be given a name. Its influence was always dreaded in judicial astrology, being unfavorable to the farmer's work. However, when rising in a geneathical chart it indicates an individual who will be loyal to his friends, guarding their secrets in faithful silence. Such a native will be a confidante to heads of state or be entrusted with public

Table 10—Named Stars of the Ploughman

Designation	Name	Type	Apparent Magnitude	RA		Decl.	
θ-Boo	Asellus	F8	4.1	14h	24m	51°	58'
μ-Boo	Alkaurops	F0	4.5	15	24	37	28
α-Boo	Arcturus	K0	0.2	14	15	19	19
γ-Boo	Haris (Seginus)	F0	3.0	14	31	38	25
ε-Boo	Mirak (Pulcherrima)	K0	2.7	14	44	27	11
β-Boo	Merez (Nekkar)	G5	3.6	15	01	40	29
η-Boo	Mufid	G0	2.8	13	53	18	31

funds, but when setting and aspected by Saturn and Mercury the native may betray his trust and end in disgrace.

Hippocrates made much of the influence of Arcturus on the human body, claiming that if the season were dry after its rising, "it will be agreeable with those who are phlegmatic and of a humid temperament, and with women; but it is most inimical to the bilious...diseases are especially apt to prove critical in these days...." The star indicates much trouble for the native in whose chart it is matutine rising or setting, but if the aspect is matutine culmination Arcturus brings riches and honor.

The Constellation of the Northern Crown

Today called Corona Borealis (CrB), its stars according to Ptolemy are of the nature of Venus and Mercury. The constellation is correlative with that which concerns the air, and especially the weather and its effects on agriculture. It is generally a beneficial constellation. When in matutine subsolar aspect the native will be fond of pleasure. When in matutine culmination, however, the native may prefer pleasure to the honors of life. If matutine setting, the constellation presages a propensity for pleasure that may even bring about disease, dishonor, and imprisonment.

Note that the two named stars follow exactly the natures as given the constellation by Ptolemy, Alphecca having the nature of Venus and Nusakan the nature of Mercury. Alpheccca is noted for conveying honor, dignity, and artistic ability. But like all Venusian stars it can also have its undesirable effects. Medieval astrologers, who tended to emphasize the more perverse side of astrology, asserted that Alphecca rising indicates a life spent in a variety of pleasurable pursuits by an individual who decked his body with adornments, secretly engaged in love affairs or adultery, and who "bedded both boys and girls," according to Firmicus Maternus. If the star is setting, disgrace will come of all this. On the more positive side this star is also indicative of those who are gifted in astrology.

Table 11—The Constellation of the Northern Crown

The Stars According to Ptolemy	Modern Designation
The bright star in the crown	α-CrB
The most western of all	β-CrB
The one east and north of this one	θ-CrB
The one again east and north of this last	π-CrB
The star east of the bright one southward	γ-CrB
The star east of this last and nearby	δ-CrB
The star still east of these	ε-CrB
The star east of all those in the crown	ξ-CrB

Table 12—Named Stars of the Northern Crown

Designation	Name	Type	Apparent Magnitude	Position RA	Decl.
α-CrB	Alphecca (Gemma)	A0	2.3	15h 34m	26° 48'
β-CrB	Nusakan	F0	3.7	15 27	29 11

The Constellation of The Man Kneeling

This constellation is said to have been an object of worship in ancient Phoenicia as the sky representative of the great sea-god Melkarth. In classical time it was the constellation of the man without a name, but today it is called Hercules (Her). *Tetrabiblos* lists its nature as that of Mercury, but Ptolemy was probably referring to the star Sarin (δ-Her) which he listed as bright. Ras Algethi has the nature of Saturn. Kajam is not listed at all by Ptolemy, but was well known to Islamic astrologer as a star astrologically as Mercury. The constellation is indicative of events that affect large portions of mankind, and the Man Kneeling presages earthquakes if poorly placed with respect to an eclipse (see Chapter VI).

Table 13—Named Stars of The Man Kneeling

Designation	Name	Type	Apparent Magnitude	Position RA	Decl.
ω-Her	Kajam	A0	4.5	16h 24m	14° 05'
β-Her	Kornephoros (Rutilicus)	K0	2.8	16 29	21 33
λ-Her	Maasym	K-0	4.5	17 30	26 08
κ-Her	Marsik	G4	5.3	16 07	17 07
α-Her	Ras Algethi	M3	3.5	17 14	14 25
δ-Her	Sarin	A2	3.2	17 14	24 52

Table 14—The Constellation of The Man Kneeling

The Stars According to Ptolemy	Modern Designation
The star in the head	α-Her
The star in the right shoulder beside the armpit	β-Her
The star in the right arm	γ-Her
The star in the right elbow	κ-Her
The star in the left shoulder	δ-Her
The star in the left arm	λ-Her
The star in the left elbow	μ-Her
The eastern one of the three in the left wrist	o-Her
The northern one of the remaining two	ν-Her
The southern one of these	ξ-Her
The star in the right side	ζ-Her
The star in the left side	ε-Her
The one north of this last in the left buttock	59-Her
The star at the beginning of the same thigh	ζ-Her
The western star of the three in the left thigh	π-Her
The one east of this last	69-Her
The star again east of this one	ρ-Her
The star in the left knee	θ-Her
The star in the let shin	ι-Her
The western one of the three in the foot	77-Her
The middle one of the three	82-Her
The eastern one of these	88-Her
The star at the beginning of the right thigh	η-Her
The one north of this and in the same thigh	σ-Her
The star in the right knee	τ-Her
The southern star of the two under the right knee	φ-Her
The northern one of these	υ–Her
The star in the right shank	χ-Her
The star at the end of the right foot which is the same as that at the tip of the crook	ν-Her
The Unfigured Star	
The star south of that in the right arm	ω-Lyr

The star Ras Algethi (α-Her) presages both the good and bad sides of Saturn. When rising it can mean a wise and clever man. But it can also indicate an individual trained in various tricks, a liar who deceives with different kinds of plots. Aspects of the horoscope (Ascendant) by Mars when Ras Algethi is rising is indicative of an aggressive personality who always displays an unbridled

Table 15—The Constellation of The Lyre

The Stars According to Ptolemy	Modern Designation
The bright star on the shell called Lyre	α-Lyr
The northern one of the two lying next to it	ε-Lyr
The southern one of these	ζ-Lyr
The star east of these and at the beginning of the shell	δ-Lyr
The northern one of the two near the eastern side of the shell	η-Lyr
The southern one of these	θ-Lyr
The northern one of the two western stars in the crossbar	ρ-Lyr
The southern one of these	ν-Lyr
The northern one of the two in the crossbar	γ-Lyr
The southern one of these	λ-Lyr

Table 16—Named Stars in The Lyre

Designation	Name	Type	Apparent Magnitude	Position RA	Decl.
η-Lyr	Aladfar	B3	4.5	19h 13m	39° 06'
β-Lyr	Sheliak	B2	3.4	18 49	33 20
γ-Lyr	Sulaphat	A0	3.3	18 58	32 39
α-Lyr	Vega	A0	0.1	18 36	38 46

hostility toward those he comes in contact with. When setting the star is indicative of a native whose life will be affected by devious plots.

The Constellation of The Lyre

The Lyre, or Lyra (Lyr), is an airy constellation that has a most marked influence on the weather, at least according to the classicists ad the astrological principles of that era. Its stars traditionally have the nature of Venus and Mercury, but Venus and Jupiter would be a better representation according to the spectra of the brightest stars in the group.

The star Vega (or Wega) is one of the brightest stars in the heavens. It is definitely of the nature of Venus, and it takes its nature on rising from the solar house of this planet, Libra. Natives in whose charts Vega is rising may be trained in the duties of the court. They will be avengers of crime and in charge of public courts and inquiries. If Saturn is in any aspect with the horoscope when Vega is rising the native will pursue this vocation with a fanaticism to the point of utilizing tortures and executions to stamp out what he considers evil. Vega was the pole star about 15,000 years ago and will be again in 12000 A.D. In classical time it was said to be a particularly good omen when Vega was matutine subsolar. In natal charts this can indicate a poetical and harmonious nature.

Table 17—The Constellation of The Bird

The Stars According to Ptolemy	Modern Designation
The star in the beak	β-Cyg
The one east of this and in the head	φ-Cyg
The star in the middle of the throat	η-Cyg
The star in the breast	γ-Cyg
The brighter star in the tail	α-Cyg
The star in the elbow of the right wing	δ-Cyg
The southern star of the three in the left wingspread	θ-Cyg
The middle one of the three	ι-Cyg
The northern one at the edge of the wingspread	κ-Cyg
The star in the elbow of the left wing	ε-Cyg
The star north of these and in the middle of the same wing	λ-Cyg
The star at the edge of the right wingspread	ζ-Cyg
The star in the left foot	o-Cyg
The star in the left knee	ω-Cyg
The western one of the two in the left foot	ρ-Cyg
The eastern one of these	π²-Cyg
The star in the nebula of the left knee	3-Cyg
The Unfigured Stars	
The northern star of the two under the right wing	τ-Cyg
The northern one of these	σ-Cyg

Table 18—Named Stars of the Bird

Designation	Name	Type	Apparent Magnitude	Position RA	Decl.
π₁-Cyg	Azlfafage	B3	4.3	21h 46m	49° 12'
β-Cyg	Albireor	K0, A0	3.2	19 30	27 54
α-Cyg	Deneb (Arided)	A2	1.3	20 41	45 11
ε-Cyg	Gienah	K0	2.6	20 45	33 53
γ-Cyg	Sador	F8	2.3	20 21	40 11
ζ-Cyg	Ruchba	K5	3.9	21 04	43 50

The Constellation of the Bird

Today called Cygnus (Cyg), in classical times it was a dreaded constellation hostile to all things of the air. Ptolemy gives its stars the natures of Venus and Mercury. This is consistent with the spectra of Deneb (Venus) and Sador (Mercury), but both Albireor and Gienah have a Martian na-

Table 19—The Constellation of Cassiopeia

The Stars According to Ptolemy	Modern Designation
The star in the head	ξ-Cas
The star in the breast	α-Cas
The star north of this and in the girdle	η-Cas
The star above the chair along the thighs	γ-Cas
The star in the knees	δ-Cas
The star in the shank	ε-Cas
The star at the tip of the foot	ι-Cas
The star in the left forearm	θ-Cas
The star below the left elbow	φ-Cas
The star in the right forearm	σ-Cas
The star above the foot of the throne	κ-Cas
The star in the middle of the back of the chair	β-Cas
The star at the end of the back	ρ-Cas

Table 20—Named Stars of Cassiopeia

Designation	Name	Type	Apparent Magnitude	Position RA	Decl.
η-Cas	Achird	F8	3.6	00 48	57° 41'
β-Cas	Caph	F5	2.4	00 08	59 01
δ-Cas	Rucha	A5	2.8	01 24	60 06
α-Cas	Schedir	K0	2.3	00 39	56 24
ε-Cas	Segin	B3	3.4	01 53	63 33

ture. Azelfafage (π_1-Cyg) is not included in the ancient catalogs. Its magnitude is 4.8 and its spectra is type B3.

When setting the star Deneb provides the native with an income from others, but Maternus asserts that badly aspected in this position Deneb presages public punishment because of the theft of the spoils of war.

The Constellation of Cassiopeia

The constellation of Cassiopeia (Cas) presages events that concern Africa and African people. Its bright stars are said to have the nature of Saturn (Schedir) and Venus (Rucha). The star Caph seems to be a trouble maker as concerns the fortunes of Islam. In 1571 there appeared in the sky next to this star a nova of such brilliance that it outshone Venus at perigee and was visible in full daylight. It was in this year that the Moslem fleet was defeated in the great battle of Lepanto. A

Table 21—The Constellation of Perseus

The Stars According to Ptolemy	Modern Designation
The nebula in the right hand	33 and 34-Per (Nebula)
The star in the right elbow	η-Per
The star in the right shoulder	γ-Per
The star in the left shoulder	θ-Per
The star in the head	τ-Per
The star in the broad of the back	ι-Per
The bright star in the right side	α-Per
The western one of the three after the one in the side	σ-Per
The middle one of the three	ψ-Per
The eastern one of these	δ-Per
The star in the left elbow	κ-Per
The bright star in Gorgon's head	β-Per
The star east of this one	w-Per
The star west of the bright one	ρ-Per
The star left farther west than this	π-Per
The star in the right knee	λ-Per
The star west of this and above the knee	43-Per
The western star of the two above the joint	μ-Per
The eastern one in the joint itself	48-Per
The star in the right calf	53-Per
The star in the right ankle	58-Per
The star in the left thigh	ν-Per
The star in the left knee	ε-Per
The star in the let shank	ξ-Per
The star in the left heel	o-Per
The star east of it in the left foot	τ-Per
The Unfigured Stars	
The star east of the one in the knee	52-Per
The star north of those in the right knee	β-Per
The western star of those in the Gorgon's head	16-Per

nova was said to be here also in 732, presaging the battle of Tours. Modern astrologers do not mention any of the named stars of this constellation. The ancients, however, believed that Schedir was a significator of makers of jewelry and those who earn their livelihood from the arts, especially as fine craftsmen.

Table 22—Named Stars in Perseus

Designation	Name	Type	Apparent Magnitude	Position RA	Decl.
β-Per	Algol (Gorgona)	B8	2-3	03h 07m	40° 52'
o-Per	Atiks	B1	3.9	03 43	32 13
ζ-Per	Menkhib	B1	2.9	03 53	31 49
η-Per	Miram	K0	3.9	02 49	55 48
α-Per	Mirfak	F5	1.9	03 23	49 47
δ-Per	Misam	K0	4.0	03 08	44 46
33 and 34-Per	al Thurayya	M	5	02 19	56 54

The Constellation of Perseus

Called Cacodaemon by medieval astrologers, Perseus (Per) contains stars of the nature of Jupiter, Saturn, Mars, and Mercury according to *Tetrabiblos*. The constellation is indicative of events affecting large numbers of people, especially those events caused by major meteorological phenomena. This constellation when prominent in a geneathical chart is said to denote adventurous individuals, but also those who are less than honest in their dealings with others.

Algol has the nature of Jupiter, but it also has a long tradition as a trouble maker. It is said to be one of the most unfortunate, violent, and dangerous stars in the heavens. This is explained by the fact that Jupiter takes on all of its most perverse meanings: too much heat, as it were, resulting in extremist actions that undo all the beneficial qualities of this planet.

Mirfak is a sign of woe in those charts in which it is prominent.

The nebula al-Thurayya is also a harmful object which portends accidents to sight and blindness. In fact, all nebulae have an evil reputation and are considered to be dangerous to the eyes.

The Constellation of The Charioteer

Now called Auriga (Aur) this is one of the most fortunate constellations in the sky, but may still portend earthquakes if situated unfortunately as regards a solar eclipse. Ptolemy lists its stars as of the nature of Mars and Mercury. This constellation culminating in a chart presages honors, especially in the field of military and political endeavors. A modern astrologer might also add sports whenever ancients mentioned military honors.

Capella, the "glorious crown," was said to be the horn of the goat that nurtured the infant Jove. The horn was broken off in play by Jove and transferred to the heavens as Cornucopiae, the "Horn of Plenty". Capella was the patron star of Babylon, and was known in Assyria as I-ku, The Leader. Capella is astrologically as the Sun, and portends civic and military honors and wealth if in matutine culmination aspect with the sun. When rising its natives will be curious about all things and have an impatient eagerness to hear anything new. But care must be taken lest the native be overly anxious and take terror at even trivial bits of new information. When setting, the curiosity of the native may lead him to reject and insult the underlying mores of the society in which he lives. The result could be the ill will of the populace, leading to death or injury from the actions of the people, or even (especially if aspected by malefics) death by public execution.

El Nath belongs to both the constellation of the Charioteer and to the constellation of the Bull (Taurus). Modern astrologers assert that it has the nature of Mars and Mercury because these planets are associated with Taurus by Ptolemy. But the spectra is that of Jupiter. This is confirmed by the fact that traditionally El Nath portends great eminence and fortune to all who could claim it as their natal star.

Hoedus I (ζ-Aur), with its double spectra, is one of the few stars with a true double nature. In this instance the star has a nature of both Jupiter and Mars. In classical times it had a bad reputation, especially among mariners who dreaded its presaging the storms on the Mediterranean. When rising Hoedus I, and its companion Hoedus II, indicates a native who may be petulant and lascivious, one who is said to be involved in depraved and vicious desires. If other factors in the chart confirm these stars the native's whole appearance to others may be that of a lie, of a person who promises one thing but hides another thing in his heart. Ancient astrologer asserted these two stars presaged cowards in battle an even suicide or other early death when setting (especially when the Descendant is also in quartile or opposition aspect to Saturn). When Mars is conjunct the Descendant as either Hoedus I or Hoedus II is setting the indication is for religious fanaticism and ruin as a result of the resultant excesses.

Menkalinan is of a Venusian nature. Robson asserts that it causes ruin, disgrace and, frequently, violent death. If so, it would be the result of an excess of pleasure seeking.

The Constellation of Serpentarius

Serpentarius is now called Ophiuchus (Oph) and was noted in classical times as the constellation of those who had the power of discovering healing herbs and skill in curing the bites of poisonous serpents. According to Greek tradition the constellation is Asclepois or Aesculapius, who was a lineal ancestor of the great physician Hippocrates. Ptolemy says that its stars have the nature of Saturn and, to some degree, Venus. Pliny said that these stars were dangerous to mankind, occasioning much mortality by poisoning. But in this regard it must be noted that the constellation of Serpentarius has been the home of a large number of comets throughout history. It would be the comets, of course, that would account for the effects noted by Pliny. In a geneathical chart the constellation denotes prudence and wisdom.

The star Ras Alhague (α-Oph) is of the nature of Venus, as is Sabik (η-Oph). But these stars have generally been associated with the excesses of this planet, portending perverted morals, mental depravity, and misfortune through women. The Islamic astrologers claimed that Ras Alhague indicated snake charmers when rising, and death from the bite of a poisonous snake if setting and

Table 23—The Constellation of the Charioteer

The Stars According to Ptolemy	Modern Designation
The southern one of the two in the head	δ-Aur
The northern one and above the head	ξ-Aur
The star in the left shoulder called Capella	α-Aur
The star in the right shoulder	β-Aur
The star in the right elbow	τ-Aur
The star in the right wrist	θ-Aur
The star in the left elbow	ε-Aur
The eastern one of the two in the left wrist called the Kids	η-Aur
The western one of these	ζ-Aur
The star in the left ankle	ι-Aur
The star common to the right ankle and the horn of the Bull	β-Aur
The star north of this one near the foot	φ-Aur
The star still north of this in the buttock	σ-Aur
The little star above the left foot	ω-Aur

Table 24—Named Stars in the Charioteer

Designation	Name	Type	Apparent Magnitude	Position RA	Decl.
α-Aur	Capella (Alhajoth)	G0	0.2	05h 15	45° 58'
β-Tau	El Nath	B8	1.8	05 25	28 35
ι-Aur	Hasseleh	K2	2.9	04 55	33 08
ζ-Aur	Hoedus I	K0, B1	3.9	05 01	41 02
η-Aur	Hoedus II	B3	3.3	05 05	41 12
β-Aur	Menkalinan	A0	2.1	05 58	44 57

aspected by a malefic. Yed Posterior (ε-Oph) was the Euphratean Nitachbat (Man of Death), and together with Yed Prior are stars of evil influence.

The Constellation of The Serpent of Serpentarius

Much of the evil influence attributed to Serpentarius is really due to this constellation, Serpens (Ser). Of stars of the nature of Mars and Saturn (the unnamed κ-Ser, for example), the constellation portends malice, deceit, and danger from poisoning. It also affects mines and minerals. The star Cor Sepentis is of the nature of Mars, but classical astrologers say it is of the nature of Saturn. In any event it is an evil mischief maker presaging accidents and violence of all kinds.

Table 25—The Constellation of Serpentarius

The Stars According to Ptolemy	Apparent Designation
The star in the head	α-Oph
The western one of the two in the right shoulder	β-Oph
The eastern one of these	γ-Oph
The western star of the two in the left shoulder	ι-Oph
The easter one of these	κ-Oph
The star in the left elbow	λ-Oph
The western star of the two in the left hand	δ-Oph
The eastern one of these	ε-Oph
The star in the right elbow	μ-Oph
The western star of the two in the right hand	ν-Oph
The eastern one of these	ζ-Oph
The star in the right knee	η-Oph
The star in the right shank	ξ-Oph
The western star of the four in the right foot	36-Oph
The one east of this	θ-Oph
The star still east of this last	b-Oph
The remaining and easternmost of the four	51-Oph
The one east of these and touching the heel	58-Oph
The star in the left knee	ζ-Oph
The northern one of the three in a straight line in the left shank	φ-Oph
The middle one of these	χ-Oph
The southern one of the three	ψ-Oph
The star in the left heel	w-Oph
The star touching the hollow of the left foot	ρ-Oph
The Unfigured Stars	
The northern one of the three east of the right shoulder	66-Oph
The middle one of the three	67-Oph
The southern one of these	68-Oph
The star east of the three above the middle one	70-Oph
The lone star north of the four	72-Oph

The Constellation of The Arrow

There is only one named star in this constellation, now called Sagitta (Sge). Ptolemy mentions that its stars are as Mars, and to some degree Venus.

Maternus asserts in his *Mathesis* that the constellation, probably meaning the star Sham, portends one who will be a successful hunter. If setting, the star indicates one who will be drafted in the army and die in battle, or who will die as a gladiator. But see below for another interpretation.

Table 26—Named Stars in Serpentarius

Designation	Name	Type	Apparent Magnitude	Position RA	Decl.
β-Oph	Kelb-Alrai	K0	2.9	17h 42m	04°35'
α-Oph	Ras Alhague	A5	2.1	17 35	12 35
η-Oph	Sabik	A2	2.6	17 09	-15 41
ε-Oph	Yed Posterior	K0	3.3	16 17	-04 38
δ-Oph	Yed Prior	M0	3.0	16 13	-03 38

Table 27—The Constellation of the Serpent of Serpentarius

The Stars According to Ptolemy	*Modern Designation*
Of the square in the head, the star at the end of the jaw	ι-Ser
The star touching the nostrils	β-Ser
The star in the temple	γ-Ser
The star at the beginning of the throat	ρ-Ser
The star in the middle of the square and in the jaw	κ-Ser
The star outside and north of the head	π-Ser
The star after the first curve of the neck	δ-Ser
The northern one of the three following this	λ-Ser
The middle one of the three	α-Ser
The southern one of these	ε-Ser
The star after the next curve, west of the Serpentarius' left hand	μ-Ser
The star east of those in the hand	υ-Ser
The star after right thigh	ν-Ser
The southern one of the two east of this	ξ-Ser
The northern one of these	ο-Ser
The star after the right hand in the tail's curve	δ-Ser
The star east of this and likewise in the tail	η-Ser
The star at the tip of the tail	θ-Ser

Table 28—Named Star in the Serpent of Serpentarius

Designation	Name	Type	Apparent Magnitude	Position RA	Decl.
θ-Ser	Alya	A5	4.5	18h 55m	+04°10'
α-Ser	Cor Serpentis	K0	2.7	15 43m	+06°30'

Table 29—The Constellation of The Arrow

The Stars According to Ptolemy	Modern Designation
The lone star in the point	γ-Sge
The eastern one of the three in the shaft	ζ-Sge
The middle one of these	δ-Sge
The western one of the three	α-Sge
The star at the extremity of the notched end	β-Sge

Table 30—Named Star in The Arrow

Designation	Name	type	Apparent Magnitude	Position RA	Decl.
α-Sge	Sham	G0	4.4	19h 39m	17° 57'

While little has been written about this constellation and its stars, we can still say a great deal by applying the basic principles of classical astrology. Its effects will certainly concern the weather, and in a geneathical chart the quality of the mind will be indicated. The tip of the Arrow is a K5 spectrum star, and the next closest one in the shaft has a double M0-A0 spectra. The tip, therefore, has the nature of Mars (Saturn) and somewhat of Venus. Here you will find violence and bodily harm, possibly as a result of jealousy.

The end of the shaft is dominated by α-Sge and β-Sge, G0 and K0 stars respectively. This combination of Mars and the Sun in a constellation such as The Arrow will result in a keen mind and a great deal of intellectual energy, but with the obvious tendency to be combative and opinionated. This same excess of the element heat will also result in rather violent weather if occurring in significant places in a chart of the new or full Moon (see Chapter VI).

Note that this lesson in the delineation of a constellation and its stars uses data that is not in the text (spectra of the unnamed stars, for example). Such data can be found in star catalogs such as *The American Ephemeris and Nautical Almanac* published by the Government Printing Office. Copies can also be found in most public libraries. Note also how the constellation itself and the position of the stars within the constellation combine to express the natures of the individual stars. It is in such a delineation that the full skill of the astrologer is required. No cookbook formulae here. The astrologer must apply basic principles such as those described in *Classical Scientific Astrology*.

The constellation of the Arrow was chosen for this analysis because of the small number of stars that it contains and because there is nothing in the literature concerning these stars. The same analysis can be made of the other constellations using both the named and unnamed stars. Such a discussion would take us too far afield, however, and so is left as an exercise for the student.

Table 31—The Constellation of The Eagle

The Stars According to Ptolemy	Modern Designation
The star in the middle of the head	τ-Aql
The star west of this and in the neck	β-Aql
The bright one in the broad of the back called The Eagle	α-Aql
The one near this to the north	γ-Aql
The western one of the two in the left shoulder	χ-Aql
The eastern one of these	π-Aql
The western one of the two in the right shoulder	μ-Aql
The eastern one of these	σ-Aql
The star farther off under The Eagle's tail touching the Milky Way	ξ-Aql
The Unfigured Stars	
The western of the two south of The Eagle's head	φ-Aql
The eastern one of these	θ-Aql
The star southwest of the Eagle's right shoulder	δ-Aql
The star south of this	ι-Aql
The star still south of this	κ-Aql
The star west of these all	λ-Aql

Table 32—Named Stars in the Eagle

Designation	Name	Type	Apparent Magnitude	Position RA	Decl.
b-Aql	Alshain	K0	3.9	19h54m	06° 21'
α-Aql	Altair	A5	0.9	19 50	08 48'
δ-Aql	Deneb Okab	K0	3.4	19 24	03 04'
γ-Aql	Tarazed (Reda)	K2	2.8	19 45	10 33

The Constellation of the Eagle

Aquila (Aql) or the Eagle portends changes in the weather and is indicative of those with a penetrating mind and clairvoyance. Ptolemy says that its stars are of the nature of Mars and Jupiter, but the brightest stars have the spectra of Mars and Venus. Altair, the name of Shaykh Ilderim's horse in *Ben Hur*, has the nature of Venus but in the Middle Ages was said to portend danger from reptiles. This probably comes from the basic Martian character of the constellation and the fact that in those days poisoning of an enemy was considered a feminine characteristic. The star Deneb Okab has a typical Martian character giving the ability to command, and success in the martial arts.

Table 33—The Constellation of The Dolphin

The Stars According to Ptolemy	Modern Designation
The western of the three stars in the tail	ε-Del
The northern one of the remaining two	ι-Del
The southern one of these	κ-Del
The southern one of the western side of those stars in the rhomboidal figure of four sides	β-Del
The northern one of the western side	α-Del
The southern one of the eastern side of the rhombus	δ-Del
The northern one of the eastern side	ν-Del
The southern one of the three between the tail and the rhombus	η-Del
The western one of the remaining two northern ones	θ-Del
The remaining eastern one of these	δ-Del

Table 34—Named Stars in The Dolphin

Designation	Name	Type	Apparent Magnitude	Position RA	Decl.
β-Del	Rotanev	F5	3.7	20h 36m	+14°30′
α-Del	Sualacin	B8	3.9	20h 39m	+15 49

The Constellation of The Dolphin

In addition to the signs of the zodiac, many of the constellations are also very important when it comes to delineating the potential character of the native at birth. Delphinus (Del) or the constellation of the Dolphin is one of these. In greece it was called the Sacred Fish. It is the sky emblem of philanthropy and its bright stars are of the nature of Mercury and Jupiter, although there are others as Saturn and Mars. It denotes an individual devoted to his children and of a very religious nature. The Martian element manifests itself in a love of hunting and sport, while that of Saturn portends a difficulty in finding true happiness. In judicial astrology the constellation presages good fortune as regards the sea.

The Constellation of the Forepart of the Horse

Now called Equuleus (Equ), this constellation was barely mentioned by Ptolemy. He did not even give the stars their astrological designation. The bright star Kitalpha has a spectral nature of Mercury and Venus combined, and Robson asserts the constellation denotes artistic ability. But ancient astrologers asserted that those born under these stars will be famous charioteers, teamsters, or courier scouts. They may also be veterinarians in keeping with the Mercurian nature of its major star.

Table 35—The Constellation of the Forepart of the Horse

The western one of the two in the head	α-Equ
The eastern one of these	β-Equ
The western one of the two in the jaw	γ-Equ
The eastern one of these	δ-Equ

Table 36—Named Stars in the Forepart of The Horse

Designation	Name	Type	Apparent Magnitude	Position RA	Decl.
α-Equ	Kitappha	F8-A3	4.1	21h 15m	05° 09'

Table 37—The Constellation of the Horse

The Stars According to Ptolemy

	Apparent
The star common to the Horse's navel and Andromeda's head	α-Peg
The star in the loin and at the end of the wing	γ-Peg
The star in the right shoulder and at the beginning of the foot	β-Peg
The star in the broad of the back, and shoulder of the wing	α-Peg
The northern one of the two in the body under the wing	τ-Peg
The southern one of these	υ-Peg
The northern one of the two in the right knee	η-Peg
The southern one of these	o-Peg
The western one of the two close together in the chest	λ-Peg
The eastern one of these	μ-Peg
The western one of the two close together in the neck	ζ-Peg
The eastern one of these	ξ-Peg
The southern one of the two in the mane	σ-Peg
The northern one of these	ρ-Peg
The northern one of the two close together in the head	θ-Peg
The southern one of these	ν-Peg
The star in the muzzle	ε-Peg
The star in the right ankle	π-Peg
The star in the left knee	ι-Peg
The star in the left ankle	κ-Peg

The Constellation of the Horse

The stars of the Horse or Pegasus (Peg) were given the nature of Mars and Mercury by Ptolemy. The constellation portends events concerning ships and the ocean, and also changes in the

Table 38—Named Stars in The Horse

Designation	Name	Type	Apparent Magnitude	Position RA	Decl.
γ-Peg	Algenib	B2	2.9	00h 12m	+15°03'
ε-Peg	Enif	K0	2.5	21 43	+09 46
τ-Peg	Kerb	A5	4.7	23 19	+23 36
a-Peg	Markab	A0	2.6	23 04	+15 04
η-Peg	Matar	G0	3.1	22 42	+30 05
β-Peg	Scheat	M0	2.6	23 03	+27 57
ζ-Peg	Homan	B8	3.6	22 40	+10 42

weather. In medieval times it was said to indicate vain individuals with a great deal of ambition, but with very poor judgment.

Algenib is of the nature of Jupiter and as such has a good reputation. However badly placed with the Sun, there will be an excess of heat resulting in violence and perhaps misfortune or dishonor.

The Martian star Enif is one that portends danger, as does Kerb. But note that while the one is a direct danger, as from being in battle, the other is an indirect danger, as from a jilted lover.

Markab portends danger from cuts or stabs and fire, but it can also indicate riches and honor.

Matar has a solar nature and was said to presage a fortunate rain.

Sheat is an especially unfortunate star regarding the sea. It was said to indicate danger from that element in the form of tidal waves or violent storms.

Of the unnamed stars ζ-Peg has a B8 spectra and is said to be very lucky, while m-Peg is a K0 star portending success in war and sports.

The Constellation of Andromeda

Tetrabiblos asserts that the bright stars of this constellation are of the nature of Venus and so is its brightest star. It is a generally favorable constellation denoting that which is honorable and eminent. Badly placed, however, it can portend earthquakes, and in the Middle Ages was said that Mars in this constellation in aspect with the Sun caused death by crucifixion or hanging.

Almach has the nature of Mars and brings with its honors in military endeavors. Some classical astrologers asserted that Almach rising in a chart portended an executioner or a jail warden, and also individuals who were foremen on construction sites and mines.

Alpheratz is of the nature of Venus and is one of the lunar stations. Being also a star in the constellation of the Horse it takes some of the nature of those stars, too. In the culminating aspects it brings riches and honor and a keen mind.

Table 39—The Constellation of Andromeda

The Stars According to Ptolemy	Modern Designation
The star common to the Horse's navel and Andromeda's head	α-And
The star in the broad of the back	δ-And
The star in the right shoulder	π-And
The star in the left shoulder	ε-And
The southern star of the three in the right arm	σ-And
The northern one of these	θ-And
The middle one of the three	ρ-Anrd
The southern star of the three at the end of the right hand	ι-And
The middle one of these	κ-And
The northern one of the three	λ-And
The star in the left arm	ζ-And
The star in the left elbow	η-And
The southern one of the three above the girdle	β-And
The middle one of these	μ-And
The northern one of the three	ν-And
The star above the left foot	γ-And
The star in the right foot	φ-And
The star south of this one	51-And
The northern one of the two in the left bend of the knee	υ-And
The southern one of these	τ-And
The star in the right knee	φ-And
The northern one of the two in the train	ξ-And
The southern one of these	ω-And
The star outside and west of the three in the right hand	o-And

Table 40—Named Stars of Andromeda

Designation	Name	Type	Apparent Magnitude	Position Ra	Decl.
η-And	Almach	K0	2.3	02h 02m	42° 13′
α-And	Alpheratz	A0	2.1	00 07	28 57
ρ-And	Mirach	M0	2.4	01 08	35 29

Mirach is of the nature of Saturn and though it brings difficulties it can also presage good fortune in marriage.

Table 41—The Constellation of The Triangle

The Stars According to Ptolemy	Modern Designation
The star at the vertex of the Triangle	α-Tri
The western of the three in the base	β-Tri
The middle one of these	δ-Tri
The eastern one of the three	γ-Tri

Table 42—Named Stars in The Triangle

Designation	Name	Type	Apparent Magnitude	Position Ra	Decl.
α-Tri	Mellellah (Mothalah)	F5	3.6	01h 52m	29° 27'

The Constellation of The Triangle

Now called Triangulum (Tri), the stars of this constellation are as Mercury according to Ptolemy. It presages events concerning rivers and streams, and also as regards the weather. It is a generally fortunate constellation, indicating a benevolent and truthful nature.

Mellellah has the planetary attribute of Mercury and indicates those of a just nature and those whose living is made in the legal profession.

Review Questions

1. θ-Aur is a relatively bright (magnitude 2.7) star in the right wrist of The Charioteer. It is spectrum A0. No name has ever been given to this star. What might its meaning be if it were rising at the time of birth? If it were culminating?

2. ζ-Oph is an unnamed star of spectrum B0 in the left knee of Serpentarius. Comment on its nature.

3. The unnamed class G5 star η-Dra is located in the body, near the tail of the Dragon. Define its nature.

4. The declination of η-Dra is 61° N 31'. Can this star rise in a chart of a native born in New York City?

5. Classical astrologers use very specific descriptions to mean very general principles such as "the native will be burned to death in his own house or by public execution" to mean that, for example, "the native's actions may lead him to be forced to depart from his home and even be a fugitive from justice (or the bill collector)." And, of course, other interpretations can be given. In this light revise what was said about Hoedus I (ζ-Aur) and Arcturu (α-Boo).

Chapter III

The Zodiacal Constellations

In discussing the constellations of the zodiac it must be remembered that much of what is said about the corresponding sign is also applicable to the asterism. Two thousand years ago the signs and the constellations were generally the same. The geometrical size of the constellations are different of course, but much of the astrological lore they portend basically is the same thing. The major effects of the constellations as opposed to the signs will be pointed out, however. These generally concern judicial astrology, which depends more on the constellations than on the sign for its delineations (see Chapter IV).

The Constellation of the Ram

Ptolemy states that the stars in the head of Aries (Ari) have an effect like the power of Mars and Saturn mingled; those in the mouth like Mercury's power and moderately like Saturn's; those in the hind foot like that of Mars; and those in the tail like that of Venus. The constellation portends events concerning sacred rites and the worship of God. It affects the conditions of the air and of the seasons, and presages the results of these elements on things that grow, especially the new shoots of arboreal crops (e.g. grapes and figs).

Botein is one of the stars in the tail, but it has a spectrum that would give it the nature of Mars.

Hamal is an unfigured star above the head of the Ram. It is of the nature of Mars and can be indicative of a violent nature.

Mesarthim and Sheratan are in the horns of the Ram. They have a spectra that would indicate a Venusian nature. Mesarthim (γ-Ari) marked the beginning of Aries, and the verbal equinox in the

Table 43—The Constellation of the Ram

The Stars According to Ptolemy	Modern Designation
The western star of the two in the horn	γ-Ari
The eastern one of these	π-Ari
The northern one of the two in the muzzle	η-Ari
The southern one of these	θ-Ari
The star in the neck	ι-Ari
The star in the loins	ν-Ari
The star at the beginning of the tail	ε-Ari
The western one of the three in the tail	δ-Ari
The middle on of the three	ξ-Ari
The eastern one of these	ζ-Ari
The star in the calf	ρ-Ari
The star under the bend of the knee	σ-Ari
The star in the hind foot	ξ-Ari
The Unfigured Stars	
The star above the head which Hipparchus placed in the muzzle	α-Ari
The eastern and brightest one of the four above the loins	41-Ari
The northern one of the remaining three	39-Ari
The middle one of the three	35-Ari
The southern one of these	33-Ari

Table 44—Named Stars of the Ram

Designation	Name	Type	Apparent Magnitude	Position RA	Decl
δ-Ari	Botein	K0	4.5	03h 10m	+19°38′
χ-Ari	Hamal	K2	2.2	02 06	+23 21
γ-Ari	Mesarthim	A0	4.7	01 52	+19 10
β-Ari	Sheratan	A5	2.7	01 53	+20 41

days of Hipparchos when the twelve signs of the zodiac were finally given their present designation.

The Constellation of the Bull

This constellation was an object of worship by primitive cultures throughout the ages. To the ancient Egyptians it was the bull-god Orissi, but according to R. A. Allen in *Star Names and Their Meanings*, the star was also worshipped by the Babylonians, Chinese, Druids, and some tribes of Amazon Indians. Its stars, according to *Tetrabiblos* are:

Table 45A—The Constellation of The Bull

The Stars According to Ptolemy	*Modern Designation*
The northern star of the four in the section	5-Tau
The one next to it	4-Tau
The one next to this last	ξ-Tau
The southernmost one of the four	o-Tau
The one east of these in the right shoulder blade	30-Tau
The star in the chest	λ-Tau
The star in the right knee	μ-Tau
The star in the right ankle	ν-Tau
The star in the left knee	90-Tau
The star in the left forearm	88-Tau
Of those in the face called the Hyades, the one on the nostrils	γ-Tau
The star between this last one and the northern eye	δ-Tau
The star between this last one and the southern eye	θ-Tau
The bright star in the Hyades in the southern eye	α-Tau
The other one in the northern eye	ε-Tau
The star at the beginning of the southern horn and the ear	97-Tau
The southern one of the two in the southern horn	104-Tau
The northern one of these	106-Tau
The star at the tip of the southern horn	ζ-Tau
The star at the beginning of the northern horn	τ-Tau
The star at the tip of the northern horn; the same as the one in the charioteer's right foot	β-Tau
The northern one of the two close together in the northern ear	ν-Tau
The southern one of these	κ-Tau
The western one of the two small ones in the neck	37-Tau
The eastern one of these	ω-Tau
The southern one of the western side of the square in the neck	41-Tau
The northern one of the western side	ψ-Tau
The southern one of the eastern side	χ-Tau
The northern one of the eastern side	φ-Tau
The northern limit of the eastern side of the Pleiades	20-Tau
The southern limit of the eastern side	17-Tau
The eastern and narrowest limit of the Pleiades	η-Tau
The small star outside and north of the Pleiades	19¹-Tau

"...like Venus along the line where the constellation is cut off [Taurus is represented as the head and fore-part only of a charging bull. These stars are 5-Tau, 4-Tau, 3-Tau, 0-Tau, and 30-Tau.]; those in the Pleiades have the nature of the Moon and Jupiter; the one in the Hyades that is bright and somewhat reddish called the Troch (Aldebaran) has a

Table 45B—The Constellation of The Bull

The Unfigured Stars

The star below the right foot and the shoulder blade	10-Tau
The western one of the three above the southern horn	ι-Tau
The middle one of the three	109-Tau
The eastern one of the these	114-Tau
The northern one of the two below he tip of the southern horn	119-Tau
The southern one of these	Σ730-Tau
The western star of the five eastern ones under the northern horn	18-Tau
The one east of this	125-Tau
The one again east of this	139-Tau
The northern one of the two remaining eastern one	136-Tau
The southern one of these	1-Gem

Table 46—Named Stars in The Bull

Designation	Name	Type	Apparent Magnitude	Position RA	Decl.
ρ-Tau	El Nath	B8	1.8	05h 25m	+28°36'
The Hyades					
γ-Tau	Hyadum I (Prima Hyadum)	K0	3.9	04 19	+15 35
δ-Tau	Hyadum II	K0	3.9	04 22	+17 30
ε-Tau	Ain	K0	3.6	04 27	+19 08
α-Tau	Aldebaran	K5	1.1	04 35	+16 28
The Pleiades					
λ-Tau	Alcyone	B5	3.0	03 46	+24 02
21-Tau	Asterope	B9	5.8	03 44	+24 26
16-Tau	Celeano	B7	5.4	03 43	+24 12
17-Tau	Electra	B5	3.8	03 44	+24 02
20-Tau	Maia	B5	4.0	03 44	+24 18
23-Tau	Merope	B5	4.2	03 45	+23 53
19-Tau	Taygeta	B5	4.4	03 44	+24 25

'temperature' like that of Mars; the others (in the Hyades) like that of Saturn and moderately like that of Mercury; and those (stars) in the tips of the horns, like that of Mars."

With the exception of El Nath (constellation of the Charioteer), all the named stars in Ptolemy's catalog of this constellation are in the Hyades or the Pleiades. The Bull is now called Taurus (Tau) and traditionally presaged the results of the beginnings of large political undertakings. The constellation also portends that which affects wild animals, especially those of danger to man.

The stars of the Hyades are α-Tau, θ-Tau, γ-Tau, δ-Tau, ζ-Tau, and ε-Tau. Neither θ-Tau nor ζ-Tau seem to have a name, however, The Hyades were the daughters of Atlas and Aethra, and half-sisters of the Pleiades, with whom they make up the fourteen Atlantides. The Hyades were the Nysean nymphs who were entrusted by Jupiter to the care and nurture of the infant Bacchus. In payment of their devotion Jupiter raised them to the sky to save them when they were driven into the sea by Lycurgus. As a group they are violent and troublesome, causing storms and tempests both on land and sea. They have always been associated with rain (even in China as well as in Greece and Rome). Edmund Spencer calls them the "Moist Daughters," and the stars have always had a favorable connotation regarding agriculture.

The most prominent star of the Hyades, indeed of the whole constellation of the Bull, is the red Aldebaran (α-Tau). Aldebaran is one of the four "Royal Stars," or "Guardians of the Sky," of Persia 5,000 years ago when it marked the vernal equinox. Before the Ram had taken the Bull's place as the first constellation of the zodiac, the star was called (in Babylonia) I-ku-u, "The Leading Star of Stars." The star is said to presage individuals who are "restless and riotous, always stirring up popular dissent and revolution." The star is also said to inflame the minds of the people with furious quarrels, and to be an enemy of quiet and peace, "madly desiring civil and domestic wars." Maternus writes that Aldebaran "...when aspected by a malefic portends sudden and unexpected involvement in riots and portends sudden and unexpected involvement in riots and sedition resulting in justly being condemned by the people." It is noteworthy that at the time of her birth Patty Hearst had Aldebaran culminating in her chart and in opposition to Mars (see Chapter V). But this star is also considered to be one of the most eminently fortunate stars in the sky, portending riches and honor (especially in war). Its natives will always seem to be under stress and in a constant state of anxiety but they are successful in making money. Like the Hyades, Aldebaran brings rain, but if the showers do not occur at its heliacal rising the prediction is for a barren year. In *The Fixed Stars and Constellations in Astrology*, Vivian Robson claims that the K0 star γ-Tau (called Primus Hyadum by modern astronomers) is the chief star of the Hyades, but this is at variance with classical literature which gives this prominence to Aldebaran.

The Pleiades are the daughters of Atlas. One day Orion saw them and became enamored and pursued them. Jupiter put them in the sky to save them from Orion. Though their number is seven, only six of the star were visible to the naked eye at the time the group was named. It is said that Electra was the missing star because she left her place so that she might not behold the ruin of Troy.

Ptolemy only mentions four of the stars of the Pleiades. We have given all the names of the seven sisters, however. In addition, two other stars of the group have been given names in more modern times: 27-Tau is named for the father of the Pleiades, Atlas, and 28-Tau for their mother, Pleione. Like the sisters, these are class B stars.

While the Pleiades have a generally good reputation during that time of the year when they are hidden by the Sun's rays (approximately forty days), they can be of great harm to mankind. At

Table 47—The Constellation of the Twins

The Stars According to Ptolemy	Modern Designation
The star in the head of the western Twin	α-Gem
The red star in the head of the eastern Twin	β-gem
The star in the left forearm of the western Twin	θ-Gem
The star in the same arm	τ-Gem
The star east of this one and in the broad of the back	ι-Gem
The one east of this in the right shoulder of the same Twin	υ-Gem
The star in the eastern shoulder of the eastern Twin	κ-Gem
The star in the right side of the western Twin	ω-Gem
The star in the left side of the eastern Twin	57-Gem
The star in the left knee of the western Twin	ε-Gem
The star under the left knee of the eastern Twin	ζ-Gem
The star in the left testicle of the eastern Twin	δ-Gem
The star under the bend of the right knee of the same Twin	λ-Gem
The star in the forward foot of the western Twin	η-Gem
The star east of this in the same foot	μ-Gem
The star at the end of the right foot of the western Twin	ν-Gem
The star at the end of the left foot of the eastern Twin	γ-Gem
The star in the right foot of the eastern Twin	ξ-Gem
The Unfigured Stars	
The western star of the forward foot of the western Twin	90-Gem
The bright western star of the western knee	κ-Aur
The western star of the left knee of the eastern Twin	36-Gem
The northern star of the three in a straight line east of the right hand of the eastern Twin	81-Gem
The middle one of the three	74-Gem
The southern one of these between the forearm and the hand	68-Gem
The bright one east of these three	ζ-Cnc

this time, with other correlative indications in the chart, they can occasion the death of a great number of people. In the charts of individuals they can presage blindness and accidents to sight at these times. Special attention should also be given to the inimical nature of the Pleiades at the time of an eclipse (see Chapter VI). But at other times the Pleiades can indeed be fortunate. They give success in agriculture and in trade conducted upon the seas. As a group they are especially indicative of events that affect the family and family life.

In a genethiacal chart the Pleiades rising indicate a person to whom luxury and lust play an important part in his or her life. They are likely to drink too much, will desire to be well dressed, and even misuse beauty aids, such as using an excessive amount of perfume. The rising Pleiades are also indicative of those who are homosexual, like to be flattered, and (with a poorly positioned

Table 48—Named Stars in The Twins

Designation	Name	Type	Apparent Magnitude	Position RA	Decl.
γ-Gem	Alhena	A0	1.9	06h 36m	+16°25'
α-Gem	Castor	A0	1.6	07 33	+31 57
ε-Gem	Mebsuta	G5	3.2	06 42	+25 09
ζ-Gem	Mekbuda	G0	3.9	07 03	+20 37
β̄-Gem	Pollux	K0	1.2	07 44	+28 05
η-Gem	Propus (Tejat Prior)	M0	3-4	06 13	+22 31
μ-Gem	Tejat (Posterior)	M0	3.2	06 21	+22 32
δ-Gem	Wasat	F0	3.5	07 19	+22 02

Mercury) impudent in speech. When setting this group of stars can have just the opposite nature. If aspected by benefics (when setting) the indication is of a pleasant death and if aspected by both malefics and benefics the native is said to be fond of the arts and perhaps even become a painter who will acquire great honors in his own lifetime. As an example of the fortunate nature of the Pleiades, Josephus, the great Jewish historian (37?-100), wrote that during the investment of Jerusalem by Antiochus Epiphanes in 170 B.C. the besieged suffered from a severe lack of water but the city was finally relieved "by a large shower of rain which fell at the setting of the Pleiades."

The most important star of this group is Alcyone. It is a class B star and of the nature of Jupiter in our scheme. It is noteworthy that all the stars of the Pleiades are also of spectral class B. Is it any wonder that Ptolemy claimed this group as having the nature of both the Moon and Jupiter?

The Constellation of The Twins

The favorable effects of Gemini (Gem), the sign of mariners and storms at sea, have already been mentioned in *Classical Scientific Astrology*. The constellation of the Twins also portends an intense devotion to others, genius, largeness of mind, goodness, and liberality. The constellation has been peculiarly connected with the fortunes of the south of England and the City of London. The Great Plague and Fire of 1665 and 1666 occurred when this constellation was in the Ascendant. The Building of London Bridge and other events of importance to the city were begun when the correlative planets were in the constellation. Chinese astrologers asserted that if this constellation were invaded by Mars, wars and a poor harvest would ensue. Ptolemy writes:

> "Of the stars in Gemini, those in the feet share the same quality as Mercury and, to a lesser degree, as Venus; the bright stars in the thighs the same as Saturn; of the two bright stars in the heads, the one in advance the same as Mercury; it is also the star called Apollo (Castor); the one in the head that follows, the same as Mars; it is also called the star of Hercules (Pollux)."

Castor is one of the Twins and Pollux the other. While Castor has a Venusian nature (according to our classification) and Pollux a nature as that of Mars, the former has always been considered to portend mischief and violence while the latter eminence and renown. This, of course, is due to the fact that the energy of Mars can be extremely valuable in such areas as war and business and politics, while Venus can cause an indolent character leading an individual to his own destruction. The other, and opposite, indications of these stars include for Castor a keen mind and many travels, and for Pollus a cruel, rash nature and a propensity to be connected with poisons.

The bright star Alhena is of the nature of Venus. It is said to give eminence in art.

The star Propus is of the nature of Saturn. It can portend eminence to those born under its influence, but such an eminence that most would not aspire to: the hangman's noose is likely to be the end result. In *The Fixed Stars and Constellations in Astrology*, Vivian Robson erroneously lists Propus as ι-Gem. This latter star is unnamed but is of spectral class K and portends success and strength.

The Constellation of the Crab

Cancer (Cnc), or the constellation of the Crab, presages thunderstorms, famine, and locusts. The stars in the eyes (according to Ptolemy) have a nature similar to that of Mercury and Mars; those in the claws to that of Saturn and Mercury; the nebula to that of Mars and the Moon; and the Asses to that of Mars and the Sun.

Acubens is spectrally as Venus, but it is in the claw and its reputation portends malevolence and poisoning.

The Asses, Asellus Australis and Asellus Borealis, portend violent death to such as come under its influence. When they get dim in the sky the indication is for rain.

The Praesepe, also known as the Beehive or the Manger, threatens mischief and blindness and is said to cause disease and disgrace. If the nebula is not visible in a clear sky it presages a violent storm. Aratos (c. 270 B.C.) wrote in *Prognostica*:

> "A murky Manger with both stars
> Shining unaltered is a sign of rain.
> If while the northern Ass is dimmed
> By vaporous shroud, he of the south gleam radiant,
> Expect a south wind; the vaporous shroud and radiance
> Exchanging stars harbinger Boreas."

As with most of classical astrology much of the effects of the planets, signs, constellations, and fixed stars can only be gleaned through direct observation of their condition in the night sky.

Table 49—The Constellation of the Crab

The Stars According to Ptolemy	Modern Designation
The middle of the nebula called The Crab in the breast	M44-Cnc
The northern star of the two western ones of the square about the nebula	η-Cnc
The southern star of the two western one	θ-Cnc
The northern star of the two eastern ones of the square called the Asses	γ-Cnc
The southern one of these two	δ-Cnc
The star in the southern claw	α-Cnc
The star in the northern claw	ι-Cnc
The star in the northern hind foot	μ-Cnc
The star in the southern hind foot	β-Cnc
The Unfigured Stars	
The star above the joint of the southern claw	π-Cnc
The star east of the tip of the southern claw	κ-Cnc
The western star of the two above the nebula	ν-Cnc
The eastern one of these	ξ-Cnc

Table 50—Named Stars of the Crab

Designation	Name	Type	Apparent Magnitude	Position RA	Decl.
α-Cnc	Acubens	A3	4.3	08h 57m	+11°57'
δ-Cnc	Asellus Australis	K0	4.2	08 43	+18 15
γ-Cnc	Asellus Borealis	A0	4.7	08 42	+21 35
M44-Cnc	Praesepe (Nubilum)	Nebula	6.0	08 38	+19 48

The Constellation of the Lion

In classical times this constellation contained the stars that today are included in the constellations of Leo and Coma Berernices (Com). *Tetrabiblos* states that the two stars in the head have the same nature as Saturn and, to a lesser degree, of Mars; the three stars in the Lion's main have the natures of Saturn and Mercury; those in the hip of the Lion and at the tip of his tail the same a Saturn and Venus; and those in the thighs the same as Venus and Mercury. The star Regulus is given the double nature of the planets Mars and Jupiter. Both the sign and the constellation are considered very fortunate regarding agriculture and Pliny asserts that the ancient Egyptians worshipped the stars of Leo because the rise of the Nile was coincident with the Sun's entrance among them. However, ancient physicians thought that when the Sun was in this constellation medicine was poison and even a bath could be harmful. It was also said that thunder from this constellation foretold sedition and the deaths of great men.

Table 51—The Constellation of the Lion

The Stars According to Ptolemy	Modern Designation
The star at the tip of the nostril	κ-Leo
The star in the open mouth	λ-Leo
The northern one of the two in the head	μ-Leo
The southern one of these	ε-Leo
The northern one of the three in the main	ζ-Leo
The middle one of the three nearby	γ-Leo
The southern one of these	η-Leo
The one at the heart called Regulus	α-Leo
The one south of this, in the chest	31-Leo
The star a little west of the one in the heart	ν-Leo
The star in the right knee	ψ-Leo
The star in the right foreclaw	ξ-Leo
The star in the left foreclaw	o-Leo
The star in the left knee	π-Leo
The star in the left armpit	ρ-Leo
The western star of the three in the belly	46-Leo
The northern one of the other eastern two	52-Leo
The southern one of these	53-Leo
The western one of the two in the loin	60-Leo
The eastern one of these	δ-Leo
The northern one of the two in the buttocks	81-Leo
The southern one of these	θ-Leo
The star in the calves of the legs	ι-Leo
The star in the hind leg joints	σ-Leo
The star south of this one, in the forelegs	τ-Leo
The star in the hind claws	υ-Leo
The star at the tip of the tail	β-Leo
The Unfigured Stars	
The western star of the two above the back	41-LMi
The eastern one of these	54-Leo
The northern one of the three above the flank	χ-Leo
The middle one of these	59-Leo
The southern one of these	58-Leo
The northernmost part of the nebula called the hair, lying between the extremities of the Lion and the Bear	75-Com
The western one of the eminent southern stars of the hair	4-Com
The eastern one of those in the figure of the ivy leaf	γ-Com

Table 52—Named Stars in The Lion

Designation	Name	Type	Apparent Magnitude	Position RA	Decl.
ζ-Leo	Andhafera	F0	3.6	10h 15m	23° 33'
γ-Leo	Algieba	K0	2.6	10 19	19 58
ε-Leo	Asad Australis	G0	3.1	09 44	23 53
μ-Leo	Asad Borelis	K0	4.1	09 51	26 08
θ-Leo	Coxa	A0	3.4	11 13	15 34
β-Leo	Denebola	A2	2.2	11 48	14 43
γ-Com	Kissen	K0	4.6	12 26	28 24
α-Leo	Regulus	B8	1.3	10 07	12 05
o-Leo	Subra	F5-A3	3.8	09 40	09 59
δ-Leo	Zosma	A3	2.6	11 13	20 38

The mercurial nature of Adafera presages storms of all kinds with thunder and lightning. It is said to indicate those whose nature is lying, stealing, and a life of crime.

In classical and medieval times a man's misfortune and disgrace was generally attributed to a woman. This perhaps explains the bad reputation of so many Venusian stars such as Denebola. However, in this instance the indications are for a noble and generous individual and the misfortune is apt to stem from this nobility of character.

The star Kissen in the Coma Berenices portends devastating rains.

The stars Asad Australis (sometimes called Algenubi) and Asad Borealis are in The Lion's head. Ptolemy gave them the nature of Saturn and Mars but the former is spectrally as the Sun. These stars are said to indicate those with an appreciation for language and a power of expression, but also those who may be heartless and cruel.

Regulus is one of the most fortunate stars in the heavens. It portends glory, riches, and power to all who are born under its influence. Proclus (412-485) called Regulus the "Royal Star," and asserted that those born under it have a royal-like nativity. In judicial astrology, however, the dimming of the star presaged evil times. A Ninevite tablet, according to R.A. Allen in *Star Names and Their Meanings*, states:

> "If the star of the great lion is gloomy the heart of the people will not rejoice."

The Venusian star Zosma indicates those who have the ability to prophesy. On the Euphrates this star, along with Coxa, was the god Kua, the Oracle. In medieval times this star was also considered to portend many of the problems connected with the planetary Venus.

Table 53—The Constellation of the Virgin

The Stars According to Ptolemy	Modern Designation
The southern star of the two in the tip of the skull	ν-Vir
The northern one of these	ξ-Vir
The northern one of the two in the face east of these	ο-Vir
The southern one of these	π-Vir
The star at the tip of the southern and left wing	β-Vir
The western of the four in the left wing	η-Vir
The one east of this	γ-Vir
The star east of this again	46-Vir
The last and eastern one of these four	θ-Vir
The star in the right side under the girdle	δ-Vir
The western star of the three in the right and northern wing	ρ-Vir
The southern of the two remaining ones	32-Vir
The northern one of these called Vindemiatrix	ε-Vir
The star in the left hand called Spica	α-Vir
The star under the girdle on the right buttock	ζ-Vir
The northern star of the western side of the square in the left thigh	74-Vir
The southern star of the western side	76-Vir
The northern star of the two of the eastern side	82-Vir
The southern one of the eastern side	β932-Vir
The star in the left knee	86-Vir
The star in the hinder part of the right thigh	90-Vir
The middle one of the three in the train about the feet	ι-Vir
The southern one of these	κ-Vir
The northern one of the three	φ-Vir
The star in the left and southern foot	λ-Vir
The star in the right and northern foot	μ-Vir
The Unfigured Stars	
The western star of the three in a straight line under the left forearm	χ-Vir
The middle one of these	ψ-Vir
The eastern one of the three	49-Vir
The western one of the three in a straight line under Spica	69-Vir
The middle one of these which is double	75-Vir
The eastern one of these	89-Vir

The Constellation of the Virgin

Now called Virgo (Vir), this constellation is indicative of an abundance in harvest and a fruitfulness of agriculture in general. But when prominent in the charts of eclipses it portends events concerning kinds (heads of state) and in this regard can be an ill omen indeed.

Table 54—Named Stars of the Virgin

Designation	Name	Type	Apparent Magnitude	Position RA	Decl.
θ-Vir	Apamni-Atsa	A0	4.5	13h09m	-05°24'
γ-Vir	Arich (Porrima)	F0	2.9	12 40	-01 19
δ-Vir	Auva (Minelauva)	M0	3.7	12 55	+03 31
ζ-Vir	Heze	A2	3.4	13 33	-00 28
α-Vir	Spica	B2	1.2	13 24	-11 03
ι-Vir	Syrma	F5	4.2	14 15	-05 53
ε-Vir	Vindemiatrix	K0	2.9	13 01	+11 06
η-Vir	Zaniah	A0	4.0	12 19	-00 32
β-Vir	Zavijah (Alaraph)	F8	3.8	11 49	+01 54

Ptolemy asserts that the stars in the head and upon the tip of the southern wing have an effect like that of Mercury and Mars, while the other bright stars of the wing and those on the girdle are like Mercury and Venus. The bright star in the northern wing, called Vindemiatrix, has the nature (according to Ptolemy) of Saturn and Mercury; its spectra, however, is that of Mars. The star Spica is of the nature of Venus and to a lesser degree that of Mars. Here again is one of the few instances where our spectral classification differs with the ancient designations. In our scheme Spica is a class B star with the nature of Jupiter. Finally, according to Ptolemy the stars in the tips of the feet and in the train have an effect like Mercury and Mars.

The star Arich indicates occult ability in the charts of the natives in which it is prominent.

Spica portends injustice to the innocent, but with a later justification and with it eminence, renown, and riches. When rising or culminating in genethliacal charts it is one of the most fortunate stars in the sky. In such a configuration it indicate rural occupations and millers of grain. When setting, however, Spica can mean imprisonment, and if aspected by Mars the enmity of the people for fraud or embezzlement of public funds.

Syrma is said to have been prominent in the charts of Mohammed and Moses. The Arabs considered this star as the most fortunate of their lunar stations.

Vindemiatrix is a mischief-making star that can indicate disgrace for the native (if male) or the untimely death of a husband (if female). Both Zaniah and Zavijah have a good astrological reputation following closely their natures as indicated by their stellar spectra.

The Constellation of The Balance

Now called Libra (Lib), this constellation in very ancient times was called "the Claws of the Scorpion." This is just another indication of the error of modern astrologers in assigning the development of the zodiac as we know it today to a time much earlier than the fifth century B.C. In-

Table 55—The Constellation of the Balance

The Stars According to Ptolemy	Modern Designation
The bright star in the southern claw of the Scorpion	α-Lib
The one north of and dimmer than this last	μ-Lib
The bright star of those at the end of the northern claw	β-Lib
The dim one west of this	δ-Lib
The star in the middle of the southern claw	ι-Lib
The star west of this in the same claw	ν-Lib
The star in the middle of the northern claw	γ-Lib
The star east of this in the same claw	θ-Lib
The Unfigured Stars	
The western star of the three northern ones in the northern claw	37-Lib
The southern star of the two eastern ones	48-Lib
The northern one of these	ξ-Sco
The eastern one of the three between the claws	λ-Lib
The northern one of the two remaining eastern ones	41-Lib
The southern one of these	κ-Lib
The western star of the more southern three of the southern claw	σ-Lib
The northern one of the two remaining eastern ones	υ-Lib
The southern one of these	τ-Lib

Table 56—Named Stars in the Balance

Designation	Name	Type	Apparent Magnitude	Position RA	Decl.
α-Lib	Zubenelgenubi	A3	2.9	14h49m	-15°56'
β-Lib	Zubeneschamali	B8	2.7	15 16	-09 18
γ-Lib	Zubenelhakrabi	K0	4.0	15 34	-14 42
δ-Lib	Zuben Elakribi	A0	var	14 59	-08 15
ν-Lib	Zuben Hakrabi	K5	5.3	15 04	-16 00
σ-Lib	Zubenalgubi	M3	3.4	15 03	-25 11

deed even Hipparchos (c. 150 B.C.) listed this constellation as being part of the constellation of the Scorpion.

The constellation has a reputation for portending the death of kings due to comets being noted here in 43 B.C. (year of Caesar's assassination) and 1106 (death of Henry IV of Germany). The constellation of Libra also indicates events concerning sacred rites and changes in religious custom. The reign of Henry IV, for example, resulted in an end of the dominance of the papacy by the kings of Germany and the emergence of the Christian church as a major secular force in Europe.

Table 57—Named Stars in the Scorpion

Designation	Name	Type	Apparent Magnitude	Position RA	Decl.
β-Sco	Acrab (Grafias)	B1	2.9	16h04m	-19°44'
α-Sco	Antares	M0, A3	1.2	16 28	-26 23
δ-Sco	Dschubba	B0	2.5	15 59	-22 33
ζ-Sco	Grafias	K5	3.7	16 53	-42 19
ν-Sco	Jabbah	B3	4.3	16 10	-19 24
υ-Sco	Lesath	B3	2.8	17 29	-37 17
λ-Sco	Shaula	B2	1.7	17 32	-37 05
κ-Sco	Girfab	B2	2.5	17 41	-39 01
θ-Sco	Sargas	F0	2.0	17 36	-42 59
σ-Sco	Al Nyat	B1	3.1	16 20	-25 32

Tetrabiblos states that the stars in the extremities of the constellation have the same influence as do Jupiter and Mercury and that those in the middle parts the same as Saturn and, to a lesser degree, Mars.

Zebeneschamali, or the Northern Scale, is considered a very fruitful star. The Chaldeans believed that when this star was clear the crops would be good. In genethiacal charts this star gives riches, honor, and good fortune.

Classical astrologers believed the same concerning the Southern Scale, Zubenelgenubi, but to a lesser degree corresponding to the difference in influence between Jupiter and Venus. Modern astrologers, however, assert just the opposite, claiming ruin and disgrace and even violence in connection with the Southern Scale.

The Constellation of the Scorpion

Scorpio (Sco) was the constellation of the dreaded symbol of darkness in the region of the Euphrates. Comets in this region of the sky were said to portend plagues of "reptiles" (read as locusts or other insects harmful to agriculture). In *Classical Scientific Astrology* the fruitful nature of Scorpio was mentioned. This attribute also applies to the constellation.

Of the named stars Antares was the "Royal Star" of the Persians. It is distinctly Martian in character, presages danger of a fatality, and makes those born on its rising rash and headstrong, but like most Martian stars it can bring honor and riches when culminating. Note also the dual spectra of Antares indicating a combination of Saturn and Venus (equals Mars). This contrasts to a nature of Mars and Jupiter postulated by Ptolemy. In fact the spectra implies a wider variation in character than does Ptolemy's designation and is more in keeping with what has been said of its effect. What is happening here is that the moisture of Venus is neutralizing the dryness of Saturn

Table 58—The Constellation of the Scorpion

The Stars According to Ptolemy	Modern Designation
The northern of the three bright ones in the forehead	β-Sco
The middle one of these	δ-Sco
The southern of the three	π-Sco
The star south of this again in one of the feet	ρ-Sco
The northern one of the two lying beside the northernmost of the bright ones	ν-Sco
The southern one of these	ω-Sco
The western star of the three bright ones in the body	σ-Sco
The middle one of these called Antares	α-Sco
The eastern one of the three	τ-Sco
The western one of the two beneath these in the farthest foot	13-Sco
The eastern one of these	d-Sco
The star in the first joint from the body	ε-Sco
The star after this one in the second joint	μ-Sco
The northern one of the double star in the third joint	ζ_1-Sco
The southern one of the double	ζ_2-Sco
The next one in the fourth joint	η-Sco
The star after this one in the fifth joint	θ-Sco
The star next to this last in the sixth joint	ι-Sco
The star in the seventh joint near the center	κ-Sco
The eastern one of the two in the center	λ-Sco
The western one of these	υ-Sco
The Unfigured Stars	
The nebula east of the center	M7-Sco
The western star of the two north of the center	45-Oph
The eastern one of these	43-Oph

so that the effect is an alternating hot and cold. Remember that these are both active elements and it is this activity that gives Antares its Martian character.

Of the other stars in the constellation Ptolemy lists the bright stars in the forehead as of the nature of Mars and Saturn; the other two stars in the body also as Mars and Jupiter; those in the joints are the same as Saturn and Venus; those in the sting of the nature of Mercury and Mars; and the nebula the same as Mars and the Moon.

Of the named stars, Dschubba and Acrab in the forehead were given the natures of Mars and Saturn by Ptolemy. However, both are a B-type star and should have the nature of Jupiter. In classical times some astrologers considered Dschubba to have the effect of Mercury. Acrab was called Grafias (or Graffias) in ancient texts. Acab is used in modern catalogs. This star is said to cause extreme malevolence, theft, crime, pestilence, and in general to be a repulsive and fiendish star.

Table 59—Named Stars in the Archer

Designation	Name	Type	Apparent Magnitude	Position RA	Decl.
π-Sgr	Albadoh	F2	3.0	19h08m	-31°04'
γ-Sgr	Al Naal	K0	3.8	18 04	-30 26
β₂-Sgr	Arkeb Posterior	F0	4.5	19 21	-44 51
β₁-Sgr	Arkeb Prior	B8	4.3	19 21	-44 30
ζ-Sgr	Ascella	A2	2.7	19 01	-29 55
ε-Sgr	Kaus Ausralis	A0	1.9	18 23	-34 24
λ-Sgr	Kaus Borealis	K0	2.9	18 26	-25 26
δ-Sgr	Kaus Medius	K0	2.8	18 19	-29 50
o-Sgr	Manubrium	K0	3.9	19 03	-21 47
σ-Sgr	Nunki (Pelegus)	B3	2.1	18 54	-26 19
μ-Sgr	Polis	B8	4.0	18 12	-24 04
α-Sgr	Rukbat	B8	4.1	19 22	-40 40

Lesath and Shaula in the sting of the Scorpion are considered unlucky stars. In medieval times these stars were said to be connected with poisons. The constellation of the Scorpion has been considered unlucky from earliest antiquity. Even though most of its bright stars have a B-type spectra they have been associated with the malevolent Mars and Saturn (read the excesses of Jupiter). This is one of the few instances of a distinct lack of correlation between what would normally be expected from the star's stellar spectra and the classical designation of its nature.

The Constellation of The Archer

Now called Sagittarius (Sgr), this constellation is the Domicile of Diana and in classical times was also known as Diannae Sidus. It is a fortunate and fruitful constellation indicating events pertaining to kings or a large portion of mankind. Ptolemy states:

> "...the stars in the point of the arrow have an effect like that of the Moon and Mars; those in the bow and the grip of the hand like that of Jupiter and Mars; the cluster in the forehead like that of the Sun and Mars; those in the cloak and the back like Jupiter, and to a lesser degree like Mercury; the stars in the feet have a nature akin to that of Jupiter and Saturn; and the quadrangle upon the tail like Venus, and to a lesser degree like Saturn."

The star in the foreankle (β-Sgr) is really two stars, but Ptolemy was probably referring only to Arkeb Prior when he designated this star.

Ascella is a fortunate star said to bring happiness.

Kaus Australis has been associated with the goddess Istar, quiet in nature with its Venus spectra. A bright nova appeared near Kaus Borealis in 386, heralding the division of the Roman Empire.

Table 60—The Constellation of the Archer

The Stars According to Ptolemy	Modern Designation
The star at the tip of the arrow	γ-Sgr
The star in the grip of the left hand	δ-Sgr
The star in the southern part of the bow	ε-Sgr
The southern star of those in the northern part of the bow	λ-Sgr
The northern of these at the tip of the bow	μ-Sgr
The star in the left shoulder	σ-Sgr
the star west of this in the arrow	φ-Sgr
The nebular and double star in the eye	M8-Sgr
The western star of the three in the head	ξ-Sgr
The middle one of these	o-Sgr
The eastern one of the three	π-Sgr
The southern one of the three in the northern part of the cloak	43-Sgr
The middle one of these	ρ-Sgr
The northern one of the three	υ-Sgr
The dim one east of these three	E-B 611-Sgr
The northern one of the two in the southern part of the cloak	55-Sgr
The southern one of these	56-Sgr
The star in the right shoulder	χ-Sgr
The star in the right elbow	52-Sgr
Of the three in the back, the star in the broad of the back	ψ-Sgr
The middle one of these in the shoulder blade	τ-Sgr
The remaining one below the armpit	ζ-Sgr
The star in the foreankle	β-Sgr
The star in the knee of the same foot	α-Sgr
The star in the right foreankle	η-Sgr
The star in the left thigh	θ-Sgr
The star in the right forearm behind	ι-Sgr
The western star of the northern side of the four at the beginning of the tail	ω-Sgr
The eastern star of the northern side	60-Sgr
The western star of the southern side	59-Sgr
The eastern star of the southern side	62-Sgr

The star Manubrium is of a Martian nature, indicating courage, heroism, and defiance.

Nunki was called the Star of the Proclamation of the Sea in the Ephratean Tablet of the Thirty Stars. Of the nature of Jupiter, this star has always portended favorable events for mariners and shipping.

Al Naal on the tip of the arrow has a nature as that of Mars. In *The Fixed Stars and Constellations*

Table 61—Named Stars in the Goat

Designation	Name	Type	Apparent magnitude	Position RA	Decl.
ε-Cap	Castra	B5	4.7	21h 36m	-19°35'
β-Cap	Dabih	G0, A0	3.2	20 20	-14 52
δ-Cap	Deneb Algedi	A5	3.0	21 46	-16 14
θ-Cap	Dorsum	A0	4.2	21 05	-17 20
α-Cap	Gredi	G5	3.8	20 17	-12 37
γ-Cap	Nashiri	F0	3.8	21 39	-16 47

in *Astrology*, Vivian Robson puts two very dim clusters and a nebula on the tip of the arrow and claims that Ptolemy mentions them as being productive of blindness. In classical times it was believed that all nebula and clusters brought on blindness. But Ptolemy does not mention any such nebula at the tip of the arrow of the Archer. Indeed, with a prominent star to mark the point, why should he?

Polis is a fortunate star portending success, ambition, and truthfulness.

The Constellation of The Goat

Capricorn (Cap), the constellation, has a great influence over human affairs portending major changes in such areas as climate and political customs. Along with the sign, the constellation is also noted as the "Mansion of Kings." Unfavorably situated with regards to lunar eclipses, it indicates major storms, especially at sea. Of the stars, *Tetrabiblos* asserts that those in the horns act in the same way as does Venus and Mars, and those in the mouth as Saturn, and, in some degree, as Venus. Those in the feet and belly act as Mars and Mercury, and those in the tail have a nature akin to that of Saturn and Jupiter.

The star Castra is of beneficial influence, although some modern astrologers such as Robson mistake this star as being in the belly and thereby give it a malevolent nature.

Dabih has a double spectra and is of the nature of the Sun and Venus combined.

Deneb Algedi is said to portend happiness and life.

Gredi, the lucky one, has a nature as the Sun and personifies the total nature of the constellation.

The star Nashire presages the bringing of good tidings. In medieval times this star was dubbed "the fortunate one."

The Constellation of the Water Bearer

The constellation of Aquarius (Aqr) is traditionally the place of good fortune. It indicates full-flowing rivers and springs and portends a sufficient amount of water for agricultural require-

Table 62—The Constellation of the Goat

The Stars According to Ptolemy	Modern Designation
The northern of the three in the eastern horn	α-Cap
The middle star of these	ν-Cap
The southern one of the three	β-Cap
The star at the tip of the western horn	ξ-Cap
The southern star of the three in the muzzle	ο-Cap
The western one of the remaining two	π-Cap
The eastern one of these	ρ-Cap
The western one of the three under the right eye	σ-Cap
The northern one of the two in the neck	τ-Cap
The southern one of these	υ-Cap
The star in the left bent knee	ω-Cap
The star under the right knee	ψ-Cap
The star in the left shoulder	24-Cap
The western star of the two close together under the belly	ζ-Cap
The eastern one of these	36-Cap
The eastern one of the three in the middle of the body	φ-Cap
The southern one of the two remaining western stars	χ-Cap
The northern one of these	η-Cap
The western one of the two in the back	θ-Cap
The eastern one of these	ι-Cap
The western one of the two in the southern part of the horn	ε-Cap
The eastern one of these	κ-Cap
The western star of the two near the tail	γ-Cap
The eastern one of these	δ-Cap
The western one of the four in the northern part of the tail	44-Cap
The southern one of the remaining three	μ-Cap
The middle one of these	λ-Cap
The northern one of these at the end of the tail	46-Cap

ments. When in matutine rising or in vespertine setting aspect with the Sun the indication is for rains (but not floods). In medieval astrology Saturn in the constellation of Aquarius meant that mankind was completely dominated by this planet. Ptolemy stated that the fixed stars in the shoulders exert an influence like that of Saturn and Mercury, as do those in the left arm and in the cloak. The stars in the thighs have a nature like that of Mercury and, to a lesser degree, of Saturn. The stars in the stream of water are, according to Ptolemy, akin to Jupiter and Saturn. While none of the named star are of the Saturnian spectra, the house of Saturn can be expected to exert its influence throughout the constellation.

Table 63A—The Constellation of the Water Bearer

The Stars According to Ptolemy	*Modern Designation*
The star in the head of the Water Bearer	25-Aqr
The brighter of the two in the right shoulder	α-Aqr
The dimmer one beneath this	σ-Aqr
The star in the left shoulder	β-Aqr
The star under the back, as if under the armpit	ξ-Aqr
The eastern one of the three in the left hand in the strap	ν-Aqr
The middle one of these	μ-Aqr
The western one of the three	ε-Aqr
The star in the right forearm	γ-Aqr
The northern one of the three in the end of the right hand	π-Aqr
The western one of the remaining two northern ones	ζ-Aqr
The eastern one of these	η-Aqr
The western star of the two close together in the right socket	θ-Aqr
The eastern one of these	ρ-Aqr
The star in the right buttock	σ-Aqr
The southern star of the two in the left buttock	ι-Aqr
The northern one of these	38-Aqr
The southern star of the two in the right shin	δ-Aqr
The northern one of these below the bend of the knee	τ-Aqr
The star in the left calf	53-Aqr
The southern one of the two in the left shin	68-Aqr
The northern one of these under the knee	66-Aqr
The western one of these in the flow of water from the hand	κ-Aqr
The star near this last southward	λ-Aqr
The star this last after the curve	83-Aqr
The star still east of this	φ-Aqr
The star in the curve south of this last	χ-Aqr
The northern one of the two south of this	ψ_1-Aqr
The southern one of the two	ψ_2-Aar
The lone star distant from these towards the south	94-Aqr
The western star of the two close together after this one	ω_2-Aqr
The eastern one of these	ω_2-Aqr
The northern one of the three in the nearby stream	103-Aqr
The middle one of the three	107-Aqr
The eastern one of these	108-Aqr
The northern one of the next three in like manner	98-Aqr
The southern one of the three	101-Aqr
The middle one of these	99-Aqr

Table 63A—The Constellation of the Water Bearer

The Stars According to Ptolemy	Modern Designation
The western one of the three in the remaining stream	86-Aqr
The southern one of the remaining two	89-Aqr
The northern one of these	88-Aqr
The last star in the water and in the southern Fish's mouth	α-PsA
The Unfigured Stars	
The western star of the three stars east of the water's curve	2-Cet
The northern one of the remaining two	6-Cet
The southern one of these	7-Cet

Table 64—Named Stars of the Water Bearer

Designation	Name	Type	Apparent Magnitude	Position RA	Decl.
ε-Aqr	Albali	A0	3.8	20h46m	-09°35'
θ-Aqr	Ancha	K0	4.3	22 16	-07 53
α-PsA	Fomalhaut	A3	1.3	22 56	-29 45
γ-Aqr	Sadachbia	A0	4.0	22 21	-01 30
α-Aqr	Sadalmelek	G0	3.2	22 04	-00 27
β-Aqr	Sadalsud	G0	3.1	21 30	-05 41
κ-Aqr	Situla	K1	5.3	22 36	-04 24
δ-Aqr	Skat (Scheat	A2	3.5	22 53	-15 57

Modern catalogs put Fomalhaut in the constellation of the Southern Fish. In classical times, however, it was considered a part of the constellation of the Water Bearer. For astrological purposes it should still be related to Aquarius. This Venusian star portends great eminence, fortune and power to those who are born under it.

Sadachbia is also called "The Lucky Star of Hidden Things." In horary astrology it indicates the ability to discover that which is hidden or lost, especially when the star is emerging from the Sun's rays (i.e. helical rising).

Both Sadalmelek and Sadalsud are of the solar spectra. Sadalmelek is the "lucky one of the King(dom)," but can portend danger and cause persecution and even death. The star Sadalsud is the "star of mighty destiny." Its solar nature presages great honor and riches. Some modern astrologers assert that Sadalsud indicates trouble and disgrace.

The stars Albali and Ancha like Sadakmelek are indicative of danger and can cause persecution and even death, but are also said to give good fortune.

Table 65—The Named Stars of the Fishes

Designation	Name	Type	Apparent Magnitude	Position Ra	Decl.
η-Psc	Al Pherg	G5	3.7	01h30m	+15°13'
α-Psc	Al Risha	A2	4.3	02 01	+02 39

Table 66A—The Constellation of the Fishes

The Stars According to Ptolemy	Modern Designation
The star in the mouth of the western Fish	β-Psc
The southern one of the two in the top of his head	γ-Psc
The northern one of these	η-Psc
The western one of the two in the back	θ-Psc
The eastern one of these	ι-Psc
The western one of the two in the belly	κ-Psc
The eastern one of these	λ-Psc
The star in the tail of the same Fish	ω-Psc
The first star from the tail, in the cord	41-Psc
The eastern one of these	51-Psc
The western star of the three bright ones following	δ-Psc
The middle one of these	ε-Psc
The eastern one of the three	ζ-Psc
The northern one of the two little ones in the curve under these	80-Psc
The southern one of these	89-Psc
The western one of the three after the curve	μ-Psc

The star Skat gives good fortune and lasting happiness and also indicates safety in a deluge.

The Constellation of the Fishes

The constellation of Pisces (Psc) portends events concerning the sea, especially those that affect the destiny of kings and large numbers of mankind. Ptolemy wrote that of the fixed stars in the constellation:

> "...those in the head of the southern fish act in the same manner as does Mercury; and those in the body (of the southern fish) as do Jupiter and Mercury; those in its tail and in its southern cord as do Saturn and in some degree as does Mercury also. The stars in the body and backbone of the northern fish have a nature akin to that of Jupiter and Saturn."

The star Al Pherb is indicative of success through determination. According to Ptolemy, Al Risha, the bright star marking the knot in the cord, is of the nature of Mars and, to some degree, Mercury. This star, however, has the A2 spectrum of Venus. It is of a unifying influence.

Table 66B—The Constellation of the Fishes

The Stars According to Ptolemy	Modern Designation
The middle one of these	ν-Psc
The eastern one of the three	ξ-Psc
The star in the knot of the two cords	α-Psc
The star west of the knot, in the northern cord	ο-Psc
The southern one of the next three after it	π-Psc
The middle one of three	η-Psc
The northern one of the three, and at the tail's end	ρ-Psc
The northern one of the two in the mouth of the eastern Fish	σ-Psc
The southern one of these	τ-Psc
The eastern of the three little ones in the head	68-Psc
The middle one of these	65-Psc
The western one of the three	67-Psc
The western star of the three in the southern fin after the star in Andromeda's elbow	ψ_1-Psc
The middle one of these	ψ_2-Psc
The eastern one of the three	ψ_3-Psc
The northern one of the two in the belly	υ-Psc
The southern one of these	φ-Psc
The star in the eastern fin under the tail	ξ-Psc
The Unfigured Stars	
The western star of the two northern ones of the square under the western Fish	27-Psc
The eastern one of these	29-Psc
The western star of the southern side	30-Psc
The eastern one of the southern side	33-Psc

Review Questions

1. How do the natures of the signs and their corresponding constellations differ?

2. What is the nature of the two unnamed stars of the Hyades (θ-Tau, class F0, and ζ-Tau, Class B3)? How do they differ from the other stars in the group? Comment.

3. What is the probable effect of the star Kissen (γ-Com) rising in a genethiacal chart?

4. θ-Sco is a bright (magnitude 2.0) star in the fifth joint from the body of the Scorpion. Its spectrum is F0. What is the effect of this star in an individual's chart?

5. The star in the northern claw of the Crab is the dim (magnitude 4.2) ι-Cnc. Its spectra is G5. What is the nature?

Chapter IV

The Southern Constellations

The southern constellations were not studied as extensively in antiquity as were the others. Some of the constellations have no named stars, and some of their brightest stars were completely unknown to the classical astrologers. But some of the most important points in the sky are in the southern constellations: Orion and the dog star Sirius, for example. These are indicated in this chapter. The following chapter will give examples of the use of the fixed stars in natal charts. The use of the constellations and the fixed stars in judicial astrology is presented in Chapter VI.

The Constellation of the Sea Monster

Now called Cetus (Cet), the stars of this constellation are said to have a nature akin to that of Saturn by Ptolemy when writing in *Tetrabiblos*. However, he was likely referring only to Menkar in this regard. Many of the brighter stars of this constellation have a Martian nature, confirming the opinion of classical astrologers that the constellation itself is correlative with the ability to command, especially in war.

Baten Kaitos is a case in point. This warlike star portends falls and blows and other misfortune brought about by force.

Deneb Kaitos (Diphda or Difda) also shows its martial nature by presaging compulsory changes. In medieval astrology this star had a bad reputation being correlative with misfortune, disgrace, and self-destruction through force.

Menkar is typical of stars with a Saturnian nature. It portends danger from beasts, disgrace, ill fortune, and illness to all those born under its influence. When rising in a genethliacal chart the

Table 67—The Constellation of the Sea Monster

The Stars According to Ptolemy	Modern Designation
The star at the nostril tip	λ-Cet
the eastern star of the three in the muzzle at the tip of the jaw	α-Cet
The middle one of these in the middle of the mouth	γ-Cet
The western one of the three in the cheek	δ-Cet
The star in the brow and eye	ν-Cet
The star north of this in the hair of the head	ζ$_2$-Cet
The star west of these in the flowing hair	ξ-Cet
The northern star of the western side of the square in the chest	ρ-Cet
The southern star of the western side	σ-Cet
the northern star of the eastern side	ε-Cet
The southern star of the eastern side	π-Cet
The middle one of the three in the body	τ-Cet
The southern one of these	υ-Cet
The northern one of the three	ζ-Cet
The eastern one of the two near the tail	θ-Cet
The western one of these	η-Cet
The northern star of the eastern side of the square near the tail	21-Cet
The southern one of the eastern side	φ$_3$-Cet
The northern star of the western side	Σ49-Cet
The southern one of the western side	φ$_1$-Cet
The northern star of the two at the tail's tip	ι-Cet
The star at the tail's southern end	β-Cet

Table 68—Named Stars in the Sea Monster

Designation	Name	Type	Apparent Magnitude	Position RA	Decl.
ζ-Cet	Baten Kaitos	K0	3.9	01h50m	-10°27'
β-Cet	Deneb Kaitos (Diphda)	K0	2.2	00 42	-18 07
γ-Cet	Kaffaljidhma	A2	3.6	02 42	+03 08
α-Cet	Menkar	M0	2.8	03 01	+03 59
o-Cet	Mira	M5	2-10	02 18	-03 05

star is said to indicate fishermen or those who make a living in some way from salt. If Menkar is aspected by Mars when setting the indication is "a violent death."

Ptolemy's catalog did not mention the variable star Mira. It is included in the list of named stars because of its potential importance when at its brightest.

The Constellation of Orion

The most important constellation in the heavens outside the zodiac is Orion (Ori). Indeed, in some instances this group of star is more significant than some of the zodiacal constellations. In judicial astrology this is the constellation of war! The Greek historian Polybios (c. 200 B.C.) attributed the loss of a Roman squadron in the first Punic War to its having sailed just after the rising of Orion. Milton in *Paradise Lost*, recounted this tragedy:

> "...when with winds Orion arm'd
> Hath vex'd the Red-sea coast, whose waves o'erthrew
> Busiris and his Memphian chivalry."

In the Middle Ages this constellation was said to be a direful influence to agriculture, being "the veri cutthrote of cattle" and presaging violent storms and much rain. In genethiacal charts the indications are for arrogance, rebellion, strength, and courage. *Tetrabiblos* lists the stars in the shoulder of Orion to have a nature similar to that of Mars and Mercury, and the other bright stars in the constellation to be similar to the natures of Jupiter and Saturn. A thousand years ago the midnight rising of this constellation indicated the season of grape-gathering in Europe.

The star Alnilam portends public honors to all those born under its influence. It was also said to indicate a good nut crop. When rising, Alnilam, together with Alnitak and Mintaka, portend irreligious and treacherous individuals. These stars, in medieval times, were said to presage those who were "passionately devoted to hunting, but not noble hunting with falcon or bow." Aspected by Saturn, the indications are for excellent fishermen, and in a female chart any of these three star rising and aspected by both Mars and a benefic portend a shrew. When setting and aspected by Mars any of the three stars indicate individuals who will incur all kinds of dangers. Mintaka, alone, portends good fortune as regards the growing of grain crops.

Bellatrix traditionally is the natal star of all those destined to great civil or military honors. It also indicates lucky and loquacious women. Thomas Hood (c. 1590) claimed that "women born under this star shall have mighty tongues."

Ensis, like other nebula, causes blindness and violent death.

Heka (λ-Ori), "the nebula in Orion's head," is not really a nebula at all, but a pale white double star. However, to an observer without a telescope it does indeed look like a nebula. Ptolemy called it the "Nebulous One," and early modern catalogs such as Flamsteed's "in capite Orionis Nebulosa" also designate it as such. Astrologically, as the Moon in our scheme, it was said to be an unfortunate influence in human affairs by the classicists. All nebulae had such a characteristic in ancient times, and were generally given a lunar nature by Ptolemy. Perhaps Heka has an undeserved reputation.

War and carnage are presaged by Betelgeuse. The star is indicative of great fortune, martial honors, and other kingly attributes. When rising or culminating the native will be a superior athlete

Table 69—The Constellation of Orion

The Stars According to Ptolemy	Modern Designation
The nebula in Orion's head	λ-Ori
The bright red star in the right shoulder	α-Ori
The star in the left shoulder	γ-Ori
The eastern star below this	32-Ori
The star in the right elbow	μ-Ori
The star in the right forearm	74-Ori
The eastern double star of the southern side of the square in the right hand	ξ-Ori
The western star of the southern side	ν-Ori
The eastern star of the northern side	72-Ori
The western star of the northern side	69-Ori
The western star of the two in the club	χ_1-Ori
The eastern one of these	χ_2-Ori
The eastern one of the four stars in a straight line in the back	ω-Ori
The star west of this one	38-Ori
The star still west of this last	33-Ori
The last and western one of the four	ψ-Ori
The northern one of those in the skin held by the left hand	15-Ori
The second one from the northernmost	11-Ori
The third from the northernmost	o_2-Ori
The fourth from the northernmost	π_1-Ori
The fifth from the northernmost	π_2-Ori
The sixth from the northernmost	π_3-Ori
The seventh from the northernmost	π_4-Ori
The eighth from the northernmost	π_5-Ori
The last and southernmost of those in the skin	π_6-Ori
The western star of the three in the velt	δ-Ori
The middle one of these	ε-Ori
The eastern one of the three	ζ-Ori
The star in the handle of the dagger	η-Ori
The northern one of the three bunched at the tip of the dagger	42-Ori
The middle one of these	θ-Ori
The southern one of the three	ι-Ori
The eastern star of the two under the tip of the handle	49-Ori
The western one of these	υ-Ori
The bright star in the left foot common with the Water	β-Ori
The northern one of these above the ball of the ankle join in the shin	τ-Ori
The star outside the left heel	29-Ori
The star under the right and eastern knee	κ-Ori

Table 70—Named Stars in Orion

Designation	Name	Type	Apparent Magnitude	Position RA	Decl.
ε-Ori	Alnilam	B0	1.7	05h35m	-01°13'
ζ-Ori	Alnitak	B0	2.0	05 39	-01 57
γ-Ori	Bellatrix	B2	1.7	05 24	+06 20
α-Ori	Betelgeuse	M0	0-1	05 54	+07 24
42-Ori	Ensis	B3	4.6	05 34	-04 51
ι-Ori	Hatsya	05	2.9	05 34	-05 56
λ-Ori	Heka	05	3.7	05 34	+09 55
δ-Ori	Mintaka	B0	2.5	05 31	-00 16
β-Ori	Rigel	B8	0.3	05 13	-08 14
κ-Ori	Saiph	B0	2.2	05 47	-09 41
μ-Ori	Tabit	F8	3.3	04 48	+06 55

being endowed with outstanding agility and speed of body. The native will have variable moods and a mind always anxious with the immediate problems of the day. When setting these anxieties may lead to a disturbed mind. Honors and titles will be given the native during his lifetime if the star is rising at this birth. If setting, these honors and titles will not come until after death.

The star Rigel gives splendor, honor, riches, and happiness to those who are born under it. It is seen that while the constellation may indicate a bad influence, such as those who are foolish and impious, the individual star portends all those who are apt to be self-reliant persons whose success will give them riches and eminence.

The Constellation of the River

Now called Eridanus (Eri), this constellation portends rain. It has variously been identified with the Nile and Tigris Rivers, and with the Garden of Eden. Modern astrologers claim that the constellation gives a love of knowledge and science, but there is no support of this contention in classical astrology. Eridanus does, however, indicate events concerned with rivers and streams. Ptolemy states that the star at the end of the river has a nature like that of Jupiter, and the other stars in the constellation have an influence akin to that of Saturn.

In modern configurations the star at the end of the river is Achernar (α-Eri). This star was not visible at the latitude of Alexandria 2,000 years ago, and was not mentioned in Ptolemy's catalog. It is, however, a class B star and so is spectrally as Jupiter.

The star at the end of the river in the ancient catalogs was Acamer (θ-Eri); spectrally it has the nature of Venus. It is said that Acamer gives success in public office or in the church. Sceptrum is also not in the constellation of Ptolemy. It is included in our list of named stars for completeness.

Table 71—The Constellation of the River

The Stars According to Ptolemy	Modern Designation
The star after the star in the foot of Orion at the beginning of the river	λ-Eri
The star north of this one in the band near Orion's shin	β-Eri
The eastern of the two after this one	ψ-Eri
The western one of these	ω-Eri
Again the eastern one of the next two	μ-Eri
The western one of these	ν-Eri
The eastern one of the three after this last	ξ-Eri
The middle one of these	o$_2$-Eri
The western one of the three	o$_1$-Eri
The eastern one of the four in the next interval	γ-Eri
The star west of this one	π-Eri
The star west of this last	δ-Eri
The western one of the four	ε-Eri
Likewise the eastern one of the four in the next interval	ζ-Eri
The star west of this last	ρ$_1$-Eri
The star again west of this	ρ$_2$-Eri
The western one of the four	η-Eri
The fist star in the bend of the River and touching the chest of the Sea Monster	τ$_1$-Eri
The star east of this	τ$_2$-Eri
The western one of the next three	τ$_3$-Eri
The middle one of the these	τ$_4$-Eri
the eastern one of the three	τ$_5$-Eri
The northern one of the western side of the four in a trapezium	τ$_6$-Eri
The southern one of the western side	τ$_7$-Eri
The western one of the eastern side	τ$_8$-Eri
The eastern one of this side and the last of the four	τ$_9$-Eri
The northern one of the two distant stars to the east and close together	υ$_1$-Eri
The southern one of these	υ$_2$-Eri
The easter one of the next tow after the turn	δ-Eri
The western one of these	41-Eri
The eastern one of the three in the next interval	γ-Eri
The middle one of these	φ-Eri
The western one of the three	η-Eri
The last and bright star of the River	θ-Eri

Table 72—Named Stars in the River

Designation	Name	type	Apparent Magnitude	Position RA	Decl.
τ-Eri	Argentenar	A3	4.2	03h01m	-23°43'
θ-Eri	Acamer	A2	3.4	02 57	-40 24
α-Eri	Achernar	B5	0.6	01 37	-57 22
η-Eri	Azha	K0	4.0	02 55	-09 00
o-Eri	Beid	F2	4.1	04 11	-06 54
β-Eri	Cursa	A3	2.9	05 07	-05 07
δ-Eri	Rana	K0	3.7	03 42	-09 51
53-Eri	Sceptrum	K0	4.0	04 37	-14 21
ν-Eri	Theemin	K0	3.9	04 34	-30 37
γ-Eri	Zaurak	K5	3.2	03 57	-13 35
ζ-Eri	Zibel	A3	4.9	03 14	-08 56

Table 73—The Constellation of the Hare

The Stars According to Ptolemy	Modern Designation
The northern star of the western side of the square down over the ears	ι-Lep
The southern one of the western side	κ-Lep
The northern one of the eastern side	ν-Lep
The southern one of the eastern side	λ-Lep
The star in the chin	κ-Lep
The star in the left forefoot	ε-Lep
The star in the middle of the body	α-Lep
The star under the body	β-Lep
The northern one of the two in the hind foot	δ-Lep
The southern one of these	γ-Lep
The star in the loin	ζ-Lep
The star at the tail's tip	η-Lep

The Constellation of the Hare

The ancients considered Orion pursuing Lepus (Lep) the hare as the Sun following the Moon. Orion is the solar constellation. Lepus is the lunar constellation. The Hindus called the Moon Cacin, or Sasanka, "Marked with the hare." Other Sanskrit and Cingalese tales mention the palace of the king of hares on the face of the Moon. The Aztecs saw on the Moon a rabbit thrown there by one of their gods, and even the Khoikhoin of South Africa and the Bantus associate the

Table 74—Named Stars of the Hare

Designation	Name	type	Apparent Magnitude	Position RA	Decl.
α-Lep	Arnab	F0	2.7	05h32m	-17°50'
β-Lep	Nihal	G0	3.0	05 27	-20 47

hare and the Moon in their worship. Astrologically, Lepus should therefore be considered to have a lunar nature. Ptolemy states that the individual stars of the constellation are of the nature of Saturn and Mercury. In medieval astrology the attributes of fecundity and a quick mind were associated with the constellation. The two named stars are of the nature of Mercury and the Sun.

When rising, Arneb portends excellent athletes. Traditionally, when aspected by Mars the sport is track; when aspected by the Moon and Mars, boxing and fencing; when aspected by Mercury, circus acrobats (jugglers) and throwing the javelin; and when aspected by Venus the native will be a pantomimist or actor of farce. (Acting was considered akin to sports in classical times.) When Arnab is aspected by both Mercury and Venus when rising the indications are that the native will acquire new arts and will pursue them quite industriously. However, if the star is rising and at this time Saturn is setting the native is said to migrate to faraway places, perhaps out of the necessity of being a fugitive. The setting star, of course, is weak and can mean failure or worse in any of the above.

The Constellation of the Dog

The ancient constellation of the Dog comprises what today are the constellations of Canis Major (CMa) and Columba (Col). There are also unfigured stars from the modern constellation of Monoceros (Mon) within the ancient boundaries of the Dog. Ptolemy states that the star Sirius has the nature of Jupiter and, to a lesser degree, that of Mars. The rest of the constellation of the Dog has the nature of Venus according to *Tetrabiblos*. Modern astrologers assert that the constellation indicates some danger from the darkness and night. Traditionally it is noted for its indication of beneficial rains.

But the astrology of Canis Major is really that of the Dog star, Sirius. Without a doubt this brightest star in the sky is one of the most important objects in the heavens, even rivaling the Sun and Moon in its influence. The Roman farmers sacrificed to Sirius a fawn-colored dog at their three festivals when the Sun began to approach Sirius (meridinal culmination). In Persia its name meant "Creator of Prosperity" and, as was mentioned in Classical Scientific Astrology, the Egyptians considered Sothis a "favorable star," especially in its matutine rising with the Sun. The Dog star is also at the vertices of the two triangles that mark the Egyptian constellation "X". This constellation of two trines is formed by the star Procyon (α-CMi), Betelgeuse (α-Ori), Naos (ζ-Pup), and Phakt (α-Col). As each of these stars is favorable in its own peculiar way, this grouping of a grand trine must be considered extremely propitious. In early astrology, wealth and

Table 75—The Constellation of the Dog

The Stars According to Ptolemy	Modern Designation
The brightest and red star in the face called the Dog	α-CMa
The star in the ears	θ-CMa
The star in the head	μ-CMa
The northern one of the two in the neck	γ-CMa
The southern one of these	ι-CMa
The star in the chest	π-CMa
The northern one of the two in the right knee	ν_3-CMa
The southern one of these	ν_2-CMa
The star at the end of the forefoot	β-CMa
The western star of the two in the left knee	ξ_1-CMa
The eastern one of these	ξ_2-CMa
The eastern star of the two in the left shoulder	o_2-CMa
The western one of these	o_1-CMa
The star at the beginning of the left thigh	δ-CMa
The star under the belly between the thighs	ε-CMa
The star in the joint of the right foot	κ-CMa
The star at the tip of the right foot	ζ-CMa
The star in the tail	η-CMa
The Unfigured Stars	
The star north of the Dog's head	δ-Mon
The southernmost one of the four in a straight line under the hind feet	θ-Col
The star north of this	κ-Col
The star still north of this	δ-Col
The last and northern star of the four	λ-CMa
The western star of the three in a straight line east of these four	μ-Col
The middle one of these	χ-Col
The eastern one of the three	γ-Col
The eastern star of the two bright ones under these	β-Col
The western star of these	α-Col
The last and southern star of the foregoing	ε-Col

renown was the happy lot of all those born under the influence of Sirius. This star is also said to give faithfulness and devotion and to make its natives important guardians and custodians to those they serve.

In later classical times this star and its constellation earned the evil reputation of so many Venusian stars. This was due in part to the ancient Egyptian practice of predicting the fortunes of the ensuing year from the heliacal rising of the star—a dim star presaging hard times ahead. The poet

Table 76—Named Stars in the Dog

Designation	Name	Type	Apparent Magnitude	Position RA	Decl.
ε-CMa	Adara	B1	1.6	06h58m	-28°56'
η-CMa	Aludra	B5	2.4	07 23	-29 16
ζ-CMa	Furud	B3	3.1	06 19	-30 03
β-CMa	Mirzam	B1	2.0	06 22	-17 57
γ-CMa	Muliphein	B5	4.1	07 03	-15 36
χ-Col	Phakt	B5	2.7	05 39	-34 05
α-CMa	Sirius	A0	-1.6	06 44	-16 41
δ-CMa	Wezen	F8	2.0	07 07	-26 21

Spencer (1552-1599) wrote in *The Shepheardes Calendar for Julye*:

> And now the Sonne hath reared up
>   His fyriefooted teme,
> Making his way between the Cuppe
>   And golden Diademe:
> The rampant Lyon hunts he fast,
>   With Dogge of noysome breath,
> Whose balefull barking brings in hast
>   Pyne, plagues, and dreery death.

Chinese astrologers asserted that Sirius portended attack from thieves, and Jewish astrologers held the star in abhorrence, probably due to the fact that their former masters worshipped it. (Leviticus xvii:7 has been interpreted to refer specifically to Sirius and Procyon.) Hippocrates made much in his *Epidemics* and *Aphorisms* of this star's power over the weather and its consequent effect on the physical well-being of mankind. Some of these theories were current in medicine in Italy even during the eighteenth century. The superstitions attributed the midsummer heat to the "dog days" of Sirius and claimed that dog bites were the result of the star's influence. A belief in the evil influence of the Dog star goes back to even before recorded history. Homer wrote:

> ... for Sirius' burning breath
> Taints the red air with favors, plagues and death.

But the more reasonable ancient meteorology claimed that a matutine setting or a vespertine rising at the full Moon preceding the winter solstice portends an uncommonly cold winter.

All the other named stars of the constellation of Canis Major except Wezen have a spectral nature of Jupiter. This conforms very well with the propitious reputation of the constellation. Wezen, with the nature of Mercury, is linked together with Phakt in astrological lore. These two stars are called "the Good Messengers," or the bringers of good news. (In some catalogs the star β-Col is

Table 77—The Constellation of Procyon

The Stars According to Ptolemy	Modern Designation
The star in the neck	β-CMi
The bright star in the hinder parts called Procyon	α-CMi

Table 78—Named Stars in Procyon

Designation	Name	Type	Apparent Magnitude	Position RA	Decl.
β-CMi	Gomeisa	B8	3.1	07h26m	+08°20'
α-CMi	Procyon	F5	0.5	07 38	+05 17

designated as Wezen. This star is much dimmer than ζ-CMa, however, and so is not included among the named stars here.)

Phakt, of course, is one of the members of the Egyptian "X". By itself it gives beneficence, hopefulness, and good fortune.

The Constellation of Procyon

The small constellation of Canis Minor (CMi) is identified chiefly with its major star, Procyon. Both the constellation and its major star portend wealth, fame, and fortune.

Ptolemy gave Procyon the nature of Mercury and, to a lesser degree, Mars. Spectrally the star is Mercurial. Like its greater neighbor, Sirius, it foretells wealth and renown, and in all classical astrology has been much regarded. When rising the star indicate the manufacture of weapons for hunting or the raising of hunting dogs. The Euphratean title for procyon is "the Water Dog," and the Arabs called it "the Bright Star of Syria." Robson, however, asserts that the title of "Water Dog" indicates that the star and constellation portends drowning. He further associates dog bites and hydrophobia with Procyon. Two examples are given purporting to associate both Sirius and Procyon with dog bites and death from rabies in *The Fixed Stars and Constellation in Astrology*. However, planetary, rather than stellar, aspects were used, and also correlations with the positions of the planets independent of the positions of the fixed star indicated the desired possibilities. Other modern astrologers, such as DeLuce in *Complete Method of Prediction*, affirm the traditional interpretation of Procyon.

The Constellation of the Argus

The ship that carried Jason and the Argonauts in search of the Golden Fleece is today composed of the constellations Carina (Car), Puppis (Pup), Pyxis (Pyx), and Vela (Vel). In classical astrology it is correlative of events concerning the sea and shipping and of rivers and springs. It also

Table 79A—The Constellation of the Argus

The Stars According to Ptolemy	Modern Designation
The western of the two in the sternpost ornament	11-Pup
The eastern one of these	ρ-Pup
The northern of the two close together above the shield in the stern	ξ-Pup
The southern one of these	o-Pup
The star west of these	μ-Pup
The bright star in the middle of the shield	κ-Pup
The western star of the three below the shield	ρ-Pup
The eastern one of these	3-Pup
The middle one of the three	1-Pup
The star in the gooseneck of the stern	23-Pup
The northern one of the two in the stern keel	49-Pup
The southern one of these	π-Pup
The northern of those in the deck of the poop	φ-Pup
The western one of the next three	δ-pup
The middle one of these	χ-Pup
The eastern one of the three	β-Pup
The bright one east of these on the deck	ζ-Pup
The western one of the two dim stars under the bright one	α-Pup
The eastern one of these	η4038-Pup
The eastern star of the two above the bright one	$η_1$-Pup
The western one of these	$η_2$-Pup
The northern star of the three in the shields near the mastholder	ω-Vel
The middle one of these	δ-Vel
The southern one of the three	ε-Vel
The northern one of the two close together under these	β-Vel
The southern one of these	ν-Vel
The southern one of the two in the middle of the mast	β-Pyx
The northern one of these	α-Pyx
The western star of the two near the tip of the mast	γ-Pyx
The eastern one of these	δ-Pyx
The star below the third and eastern shield	λ-Vel
The star on the section of the deck	ψ-Vel
The star between the oars in the keel	σ-Pup
The dim star west of this	ρ-Pup
The bright star east of this below the deck	γ-Vel
The bright star south of this in the lower keel	χ-Car
The western star of the three east of this	ε-Car

Table 79B—The Constellation of the Argus

The Stars According to Ptolemy	Modern Designation
The middle one of these	δ-Vel
The eastern one of the three	ι-Car
Of the two east of these, the western one near the section	κ-Vel
The eastern one of these	φ-Vel
The western one of the two in the northern and western oar	η-Col
The eastern one of these	ν-Pup
Of the two in the remaining oar the western star called Canopus	α-Car
The last and eastern one of these	τ-Car

Table 80—Named Stars of the Argus

Destination	Name	Type	Apparent Magnitude	Position RA	Decl.
γ-Pyx	Alphart	K2	4.2	08h49m	27S27
λ-Vel	Alsuhail	K5	2.2	09 07	43S20
ξ-Pup	Azmidiske	G0	3.5	07 48	24S48
α-Car	Canopus	F0	-0.9	06 23	52S41
κ-Pup	Markeb	B8	4.5	07 38	26S45
β-Car	Miaplacidus	A0	1.8	09 13	69S38
ζ-Pup	Naos	0d	2.3	08 03	39S56
ι-Car	Tureis	F0	2.2	09 17	59S11

has been said to be connected with death by drowning, especially if connected with the eighth house. *Tetrabiblos* asserts that its bright stars are of the nature of Saturn and Jupiter.

The brightest star in the constellation is Canopus. It is named for the chief pilot of the fleet of Menelaos. On his return from the destruction of Troy in 1183 B.C., Canopus died. In his honor a city was founded and named after him. The ancient city of Canopus is now in ruins. Its site is occupied by the village of Al Bekur and marks the location where Ptolemy made his observations. But the star was worshipped on the Nile as early as 6400 B.C. The temples at Edfu, Philae, Amada, and Semneh are so oriented that they herald the rising of Canopus at the autumnal equinox on that date. Astrologically it is correlated with the sea in the same manner as is the constellation. The Hindus called it Agastya, a son of Varsuna, the goddess of the waters. It promises immunity from disease and indicates unrequited love when connected with Venus. It also gives piety and changes evil to good. Since the sixth century it has been the "Star of Saint Catherine," appearing to the Greek and Russian pilgrims as they approach her convent at Sinai.

The star Markeb asserts its Jupiterian character by portending profitable journeys.

Table 81—The Constellation of the Water Snake

The Stars According to Ptolemy	Modern Designation
Of the two western stars in the head, the southern one in the nostrils	σ-Hya
The northern one of these, above the eye	δ-Hya
Of the two stars east of these, the northern one near the top of the head	ε-Hya
The southern one of these, in the open mouth	η-Hya
The star east of them all in the jaw	ρ-Hya
the western one of the two at the beginning of the neck	ζ-Hya
the eastern one of these	θ-Hya
The middle one of the next three in the curve of the neck	τ₂-Hya
The eastern one of the three	ι-Hya
The southernmost of these	τ₁-Hya
The dim northern star of the two close together in the south	28-Hya
The bright one of these two	α-Hya
The western star of the three east of the curve	κ-Hya
The middle one of these	ν₁-Hya
The eastern one of the three	ν₂-Hya
The western one of the next three almost in a straight line	μ-Hya
The middle one of these	φ-Hya
The eastern one o the three	υ-Hya
The northern one of the two after the base of the bowl	β-Hya
The southern one of these	χ-Hya
The western one of the three in the triangle after these	ξ-Hya
The middle one south of these	ο-Hya
The eastern one of the three	β-Hya
The star behind the raven near the tail	γ-Hya
The star at the tip of the tail	π-Hya
The Unfigured Stars	
The star in the southern part of the head	c-Hya
The star far east of those in the neck	-Hya

The southeastern terminal of the Egyptian "X" is marked by Naos. This star, with its lunar spectra is as favorable as is its planetary counterpart.

The Constellation of the Water Snake

Today called Hydra (Hya), this constellation is similar in nature to its northern counterpart, Draco. This group of stars also affects the sea and shipping, however. In this latter instance the indications are for difficulties and even tragedy. Ptolemy states that the bright stars of the constellation have the natures of Saturn and Venus.

Table 82—Named Stars in the Water Snake

Designation	Name	Type	Apparent Magnitude	Position RA	Decl.
α-Hya	Cor Hydrae (Alfard)	K2	2.2	09h26m	-08°31'

Table 83—The Constellation of the Bowl

The Stars According to Ptolemy	Modern Designation
The star in the base of the bowl common to the Water Snake	α-Crt
The southern one of the two in the middle of the Bowl	γ-Crt
The northern one of these	δ-Crt
The star on the southern edge of the face	ζ-Crt
The star on the northern edge	ξ-Crt
The star on the southern handle	η-Crt
The star on the northern handle	θ-Crt

Table 84—Named Stars in the Bowl

Designation	Name	Type	Apparent Magnitude	Position RA	Decl.
α-Crt	Alkes	K0	4.2	10h59m	-18°10'

The only named star in the group is Alphard. It is said to give wisdom and knowledge of human nature. In the Middle Ages it portended immorality, revolting deeds, and a sudden death.

The Constellation of the Bowl

The only named star in the Bowl, or Crater (Crt), was given the nature of Venus and, to a lesser degree, Mercury, by Ptolemy. The constellation gives good mental abilities, but there can also be sudden changes and unexpected events.

Alkes has a spectra of Mars, and has always portended eminence to those born under its influence. When rising the star indicates dedicated environmentalists whose love for rivers and streams lead them to be very protective regarding water resources. At a less intense level the native may become a landscape architect, a builder of canals, or in some other manner do business in merchandise connected with the water.

The Constellation of the Raven

Today called Corvus (Crv), this constellation has all the traditional attributes of Mercury, but with some distinct perversions! The sharp intelligence is blunted, thoughts and logic become muddled, the memory dims, the ability to communicate is weakened, and what passes for wisdom is superficial.

Table 85—The Constellation of the Raven

The Stars According to Ptolemy	Modern Designation
The star in the beak and common to the Water Snake	α-Crv
The star in the neck near the head	ε-Crv
The star in the breast	ζ-Crv
The star in the western and right wing	γ-Crv
The western star of the two in the eastern wing	δ-Crv
The eastern one of these	η-Crv
The star at the end of the foot, common with the Water Snake	β-Crv

Table 86—Named Stars in the Raven

Designation	Name	Type	Apparent Magnitude	Position RA	Decl.
α-Crv	Alchita (Alchiba)	F2	4.2	12h07m	-24°35'
δ-Crv	Algorab	A0	3.1	12 29	-16 23
β-Crv	Kraz	G5	2.8	12 33	-23 16
ε-Crv	Minkar	K0	3.2	12 09	-22 30

Mercury is there, but any attempt to manifest its abilities ends in disasters. Ovid explained the phenomena by claiming that Apollo punished the Raven for her unfaithfulness. Ptolemy asserts that the stars of the constellation portend storms, and some astrologers say it gives craftiness and greediness.

The star Algorab indicates malevolence, repulsiveness, and lying.

The Constellation of the Centaur

The ancient constellation of Centaurus (Cen) also includes the stars that today are put into the constellation of Crux (Cru). Ptolemy states that the stars representing the human part of the constellation have a nature like that of Venus and Mercury, and the bright stars in the equine part of the asterism (mostly those stars in Crux) are like Venus and Jupiter. Astrologically the constellation gives wisdom and a proficiency in botany, music, astronomy, divination, and medicine. In Medieval times the constellation was said to be connected with poison and to indicate strong passions and vengeance.

The star Agena presages friendship, health, and honor. It is also said to indicate those in high positions.

Toliam (alpha-Centauri) is the closest star to our solar system, except the Sun. Astrologically it is as the Sun and it portends friends, refinement, and positions of honor. When rising, this star indicates a breeder or trainer of horses. If favorably aspected by Mars the native in less mechanistic

Table 87—The Constellation of the Centaur

The Stars According to Ptolemy	Modern Designation
The southernmost of the four in the head	γ-Cen
The northernmost of these	η-Cen
The western star of the two remaining middle ones	ι-Cen
The eastern and last one of these four	κ-Cen
The star on the left and western shoulder	ι-Cen
The star on the right shoulder	θ-Cen
The star on the left shoulder blade	δ-Cen
The northern star of the two western ones of the four in the wand	ψ-Cen
The southern one of these	α-Cen
Of the remaining two, the star at the tip of the wand	χ-Cen
The remaining one south of this	β-Cen
The western one of the three in the right side	ν-Cen
The middle one of these	μ-Cen
The eastern one of the three	φ-Cen
The star in the right arm	χ-Cen
The star in the right forearm	η-Cen
The star at the tip of the right hand	κ-Cen
The bright star at the beginning of the human body	ζ-Cen
The eastern star of the two dim stars north of this one	$υ_2$-Cen
The western one of these	$υ_1$-Cen
The star at the beginning of the back	w-Cen
The star west of this last on the horse's back	ξ-Cen
The eastern star of the three in the loins	γ-Cen
The middle one of these	τ-Cen
The western one of the three	σ-Cen
The western star of the two close together in the right thigh	δ-Cen
The eastern one of these	ρ-Cen
The star in the chest under the horse's armpit	ε-Cen
The western star of the two under the belly	δ-Cru
The eastern one of these	γ-Cru
The star in the bend of the right foot	α-Cru
The star in the ankle of the same foot	ζ-Cru
The star under the ankle of the left foot	β-Cru
The star in the front of the same foot	θ-Cru
The star at the tip of the right forefoot	α-Cen
The star in the knee of the left foot	β-Cen
The star outside under the right hind foot	β-Cru

Table 88—Named Stars of the Centaur

Designation	Name	Type	Apparent Magnitude	Position RA	Decl.
α-Cru	Acrux	B1	1.6	12h25m	-62°58′
β-Cen	Agena	B1	0.9	14 02	-60 15
α-Cen	Toliman	G0	0.1	14 38	-60 44

Table 89—The Constellation of the Censer

The Stars According to Ptolemy	Modern Designation
The northern star of the two in the base	σ-Ara
The southern one of these	θ-Ara
The star in the middle of the altar-like vessel	α-Ara
The northern one of the three in the brazier	ε-Ara
The southern one of the remaining two contiguous ones	γ-Ara
The northern one of these	β-Ara
The star at the end of the burning	ζ-Ara

Table 90—Named Star in the Censer

Designation	Name	Type	Apparent Magnitude	Position RA	Decl.
α-Ara	Censer	B3	3.0	17h30m	-49°52′

times would have been a calvary soldier, but today a jockey, polo player, or other profession requiring horsemanship would be more logical. An aspect by Venus of Toliman when rising is indicative of natives who become veterinarians. When setting, of course, the star is indicative of troubles pertaining to the above descriptions. auch problems might include horse-related accidents or falling from high places.

The Constellation of the Censer

Today called Ara, Ptolemy states that its stars are like Venus and, to a lesser degree, like Mercury. This asterism is said to give aptitude in science, a devotion to God, and success in ecclesiastical matters.

The star α-Ara is called the Censer. When rising this star was said by Maternus in *Mathesis* to indicate priests, prophets, theologians, and others associated with religion.

Table 91—The Constellation of the Wild Beast

The Stars According to Ptolemy	Modern Designation
The star at the end of the hind foot near the Centaur's hand	β-Lup
The star in the hand of the same foot	α-Lup
The western star of the two in the shoulder blade	δ-Lup
The eastern one of these	γ-Lup
The star in the middle of the Wild Beast's body	ε-Lup
The star in the belly under the flank	λ-Lup
The star in the thigh	π-Lup
The northern one of the two near the beginning of the thigh	μ-Lup
The southern one of these	k-Lup
The star at the end of the loins	ζ-Lup
The southern star of the three at the tip of the tail	ι-Lup
The middle one of the three	τ-Lup
The northern one of these	ρ-Lup
The southern star of the two in the neck	ζ-Lup
The northern one of these	θ-Lup
The western one of the two in the muzzle	χ-Lup
The eastern one of these	ξ-Lup
The southern one of the two in the forefoot	φ$_2$-Lup
The northern one of these	φ$_1$-Lup

Table 92—Named Stars in the Wild Beast

Designation	Name	Type	Apparent Magnitude	Position RA	Decl.
α-Lup	Lupus	B2	2.9	14h40m	-47°17'
β-Lup	Ke Kwan	B2	2.8	14 57	-43 02

The Constellation of the Wild Beast

Today called Lupus (Lup), the Wolf, the ancients merely called this constellation the "Wild Beast." *Tetrabiblos* asserts that its stars are like Saturn and Mars. It indicates a treacherous nature and portends strong passions and aggressions. The named stars of this constellation have no ancient astrological lore.

The Constellation of the Southern Crown

According to Ptolemy, the stars of the Corona Australis (CrA) have a nature akin to that of Saturn and Mercury (not Saturn and Jupiter as some write). The constellation is said to bring unforeseen

Table 93—The Constellation of the Southern Crown

The Stars According to Ptolemy	Modern Designation
The outside western star of the southern edge	ε-Tel
The eastern star of those on the Crown	α-Tel
The star east of this one	η-CrA
The star again east of this one	ζ-CrA
The star after this one in front of the Archer's groin	δ-CrA
The star after this one and north of the bright one in the knee	β-CrA
The star north of this one	α-CrA
The star still north of this one	γ-CrA
The eastern one of the two western ones after this last on the northern edge	ε-CrA
The western one of these two dim ones	μ-CrA
The star rather west of this one	λ-CrA
The star still west of this	κ-CrA
The last one south of this last	θ-CrA

Table 94—Named Star in the Southern Crown

Designation	Name	Type	Apparent Magnitude	Position RA	Decl.
α-CrA	Southern Crown	A2	4.1	19h 08m	-37°56'

troubles to those in a position of authority. Note that two of the stars of this group are in the modern constellation of Telescopium (Tel). There is no indication that this modern group was wholly contained in the ancient Southern Crown, however. There are no formally named stars in the group. The star α-CrA is designated as the Southern Crown is in Table 94.

The Constellation of the Southern Fish

This last of Ptolemy's forty-eight constellations, Pisces Australis (PsA), also contains the stars from the constellation of Indus (Ind), Gruis (Gru) and Microscopium (Mic). In the Middle Ages its stars were said to have the nature of Saturn. The brightest star given to this constellation in modern times is Fomalhaut. Fomalhaut was considered to be a member of Aquarius in classical times and should still be considered so for astrological purposes. The only other named star of the Southern Fish is Alnair with the nature of Jupiter. But as this star is unfigured, its influence is to be considered minimal. The constellation itself has an effect like its northern namesake, Pisces. In judicial astrology, and when related to an eclipse, this constellation also indicates events concerning the sea.

Table 95—the Constellation of the Southern Fish

The Stars According to Ptolemy	Modern Designation
The star in the mouth, the same as that at the beginning of the Water	α-PsA
The western star of the three at the southern edge of the head	β-PsA
The middle one of these	γ-PsA
The eastern one of the three	δ-PsA
The star near the gills	ε-PsA
The star in the southern spinal fin	μ-PsA
The eastern one of the two in the belly	ζ-PsA
The western one of these	λ-PsA
The eastern one of the three in the northern fin	η-PsA
The middle one of these	θ-PsA
The western one of the three	ι-PsA
The star at the tip of the tail	γ-Cru
The Unfigured Stars	
The western one of the three bright stars west of the Fish	α-Ind
The middle one of these	ζ-Ind
The eastern one of the three	α-Gru
The dim star west of this	ζ-Mic
The southern star of the remaining two to the north	α-Mic
The northern one of these	γ-Mic

Table 96—Named Stars in the Southern Fish

Designation	Name	Type	Apparent Magnitude	RA	Position Decl.
α-Gru	Alnair	B5	2.2	22h07m	-47°05′

The Zodiacal Stars

In general the planetary aspects are not applicable when considering the fixed stars. The absurdity of attempting to use such aspects with fixed stars is nowhere better illustrated than by considering the star Dziban (ψ-Dra). Transiting its position to the ecliptic this star is found to have a longitude of 12 Capricorn 55. Using the naive methodology of modern astrology, a planet at 12 Capricorn 55 would then be said to be in conjunction with Dziban. But the latitude of the star is 84°11′, a distance of almost ninety degrees. Is the aspect a conjunction or quartile? (But note that planets in aspect with the angles when a star is rising, setting, or culminating are said to aspect the star.)

But star within five degrees of the ecliptic, so-called zodiacal stars, can validly be combined with the planets in the traditional aspects. These stars may also be occulted by the planets and by the Sun and Moon. Only forty-three named stars are zodiacal.

Table 97—The Zodiacal Stars

Ain	ε-tau	Mekbuda	ζ-Gem
Alcyone	η-Tau	Merope	23-Tau
Acrab	β-Sco	Manubrium	o-Sgr
Antares	α-Sco	Nashira	γ-Cap
Apamic-Atsa	o-Vir	Nunki	σ-Sgr
Ancha	θ-Aqr	Pleione	28-Tau
Arich	γ-Vir	Polis	μ-Sgr
Asellus Australis	δ-Cnc	Praesepe	M44-Cnc
Asellus Borelis	γ-Cnc	Propus	η-Gem
Asterope	21-Tau	Regulus	α-Leo
Atlas	27-Tau	Situla	γ-Aqr
Botein	δ-Ari	Spica	α-Vir
Celeano	16-Tau	Subra	o-Leo
Dabik	β-Cap	Taygeta	19-Tau
Deneb Algedi	δ-Cap	Tejat Posterior	μ-Gem
Dorsum	θ-Cap	Wasat	δ-Gem
Dschubba	δ-Sco	Zaniah	η-Vir
Electra	17-Tau	Zavijah	β-Vir
Jabbah	ν-Sco	Zubenelgenubi	α-Lib
Kaus Borealis	λ-Sgr	Zuben Elakrab	γ-Lib
Maia	20-Tau	Zuben Hakrabi	ν-Lib
Mebsuta	ε-Gem		

Review Questions

1. Review what was said under the Constellation of the Arrow (Chapter II) and also read Appendix B.

2. The stars β-Cru and γ-Cru are very bright (magnitude 1.5 and 1.6 respectively). They are unnamed. β-Cru has a B1 spectrum and γ-Cru an M3 spectrum. Describe their astrological significance.

Chapter V

Genethliacal Applications of the Fixed Stars

The Stars One Is Born Under

By far the most important applications of the fixed stars and constellations in classical astrology occur in the field of judicial, or mundane, astrology. These applications will be discussed in detail in the next chapter. But to fix the principles that have been presented in the previous chapters, two illustrations from genethliacal astrology will be given.

Almost everyone knows the sign of the zodiac within which was posited the Sun on the day of their birth. Modern astrologers place a great deal of emphasis on this sign, and its meaning and daily influence is widely distributed in newspapers. In ancient times the stars that were rising, culminating, and setting at the time of birth were considered to be of primary importance—rather than the Sun sign. A thousand years ago everyone could recite the stars they were born under and discourse on the meanings in the same manner that people today talk of their Sun sign.

Unlike the Sun sign, which merely indicates the psychological nature of an individual, the stars one is born under presage a general outline of what a person's life may be like. Each of the stars in the heavens has the astrological nature of one of the planets; hence the effect of the stars one is born under is as if those planets were directly on the angles of the chart. For this reason classical astrologers used the fixed stars extensively in their delineations.

An individual's Sun sign is specific only to the thirty-day period of birth. The stars one is born under, on the other hand, are specific to the exact time and location of birth. That is, one's Sun sign is also applicable to millions of other people born within the same thirty-day period. In contrast, one's astrological chart, depicting the locations of all the heavenly bodies at the time of birth, is as unique to a person as fingerprints. The stars one is born under fall in between these two extremes.

The stars one is born under are determined from the local sidereal time of birth. A star whose right ascension is the same as this sidereal time will culminate at the time and place of the native's birth. Appendix C gives a formula for determining the sidereal time for rising and setting of a star. When this time corresponds to the sidereal time of birth the star will be rising or setting. Appendix C also gives tables of rising, setting, and culminating times of some of the fixed stars.

Also of interest are those stars in stellar aspect to the Sun (Chapter I). The sidereal time that the Sun rises, sets, and culminates on the day of the native's birth (rather than the time of birth) is the determining factor. The fixed stars that are rising, setting, or culminating at this time are the ones in aspect.

For those stars rising, setting, or culminating at the same time of birth an orb of fifteen minutes is used since it is rare that greater precision in the time of birth can be obtained. An orb of four minutes is used for stellar aspects as this represents one day's movement by the Sun. It may happen that more than one star is within an orb. Following the philosophy of classical astrology, the brighter star is generally chosen, but judgement is important. A very bright star that is just within orb would not be chosen over a slightly dimmer one that is exact as to time. It is also possible that a native may have more than one star rising, setting, or culminating at the time in question.

Gerald Ford

Consider the chart of former President of the United States Gerald Ford (Figure 1). He was born at 00:43 hours Central Standard Time on July 14, 1913 at Omaha, Nebraska. The local sidereal time of President Ford's birth is 19h45m, and the Sun's position of 12 Cancer 21 translates to a right ascension of 06h54m. The stars rising and culminating at the local sidereal time of birth are the primary stars under which Ford was born. The setting star at this time is also highly indicative of the native's potential at birth.

The Sun's right ascension is also the sidereal time at which the Sun will culminate on the day of Ford's birth. Therefore, stars with this same right ascension will be in a meridinal culmination aspect to his Sun. Stars that rise or set at a sidereal time of 06h54m will be in a meridinal subsolar and meridinal setting aspect, respectively. From the equations given in Appendix C it is seen that on July 14, 1913 at Omaha, Nebraska the Sun rises at a sidereal time of 23h27m and sets at a sidereal time of 14h21m. These sidereal times define the matutine and vespertine aspects.

Using the formula in Appendix C the sidereal times of rising, culmination, and setting of all the named stars are investigated to determine which of them are of interest. At 19h45m local sidereal time, it is found that the star Hoedus II (η-Aur) is rising, Tarazed (γ-Aql) is culminating, and

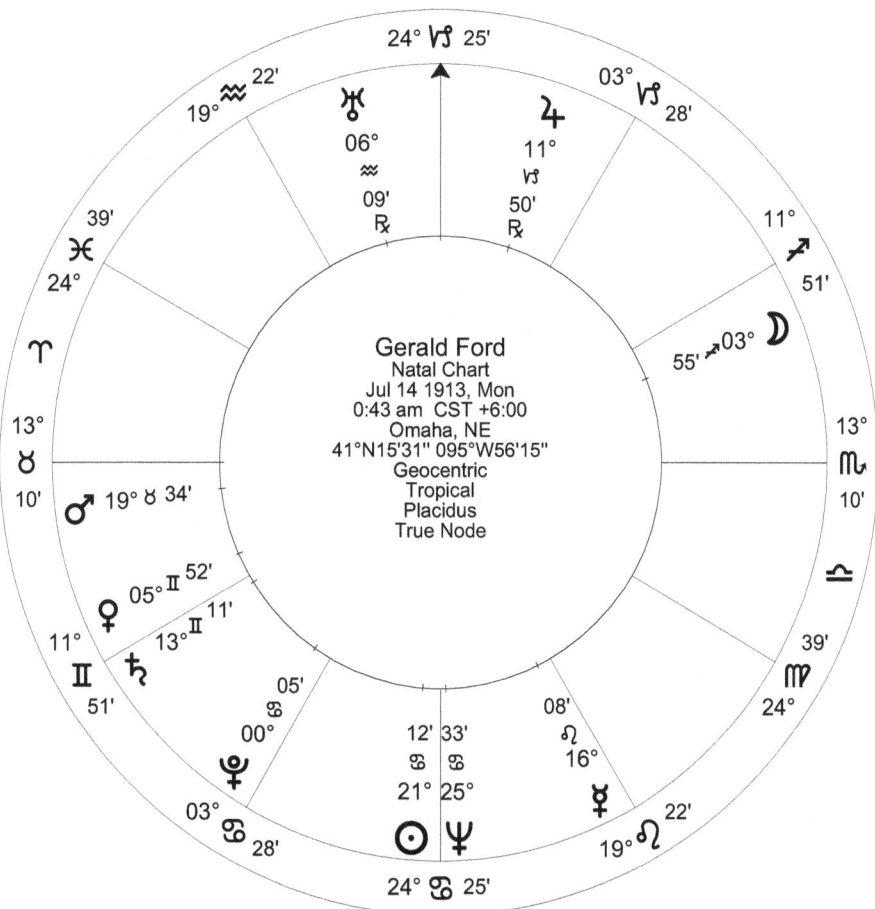

Vindematrix (ε-Vir) is setting. The locations of these stars are translated to the ecliptic and placed in the chart as noted in Figure 1. It is seen that Hoedus II has a celestial longitude of 19 Gemini 08, but the Ascendant is at 13 Taurus 08. This places the star almost thirty degrees below the Ascendant on the chart, but if one were to observe the sky at the instant of Ford's birth the star Hoedus II would be found directly on the horoscopic point! The reason for this is that the declination of the star, 41°12′, places it far above the ecliptic. It will, therefore, rise sooner than a star or planet on the ecliptic. This is a graphic example for not using longitudes when considering aspects to the fixed stars.

The star Castor (α-Gem) is matutine subsolar and Nusakan (b-CrB) is matutine setting. At the meridinal culmination is Adara (ε-Cma). Two stars are at vespertine rising aspect: Kaus Medius (δ-Sgr) and Manubrium (o-Sgr). Finally the star Mekbuda (ζ-Gem) is at the vespertine setting aspect. These stars are placed on the chart as in Figure 1. In this chart the first Greek letter of each element of the aspects is used to designate that aspect: i.e. p α indicates the matutine subsolar aspect and o λ the vespertine setting aspect.

The rising constellation Augriga indicates political honors, something Ford has achieved. The culminating constellation of Aquila brings a penetrating mind. This is confirmed by the rising star Castor which is also said to indicate a keen mind and many travels. Hoedus II, on the eastern horizon, is of the nature of Jupiter and portends that expansive personality so necessary to a politicians. On the Midheaven Tarezed has the nature of Mars, giving that energy and courage required of a leader. This is also confirmed by Manubrium (vespertine rising) and Mars in the first locus (house). But Vindematrix is setting at the time of birth and Nusakan is at the matutine setting aspect. Both these stars presage trouble: in the Middle Ages it was said that they brought disgrace. The worst of these stars is Vindematrix. But this star is setting, and stars setting at birth are in a weakened condition. Hence, rather than disgrace the prediction is for the native not to reach the fulfillment of his desire. Ford, of course, became president due to the resignation of Richard Nixon, but was unable to be elected to that office on his own.

Patty Hearst

She is the granddaughter of William Randolph Hearst, noted newspaper publisher. When in college at the University of California at Berkeley, she was kidnapped by a group of revolutionary terrorists calling themselves the Symbionese Liberation Army. They demanded a ransom, which was paid, but Hearst was not released. Instead she changed her name to Tanya and joined the revolutionaries. Until captured she was a leader in the movement, directing her "army" in a series of bank robberies and revolutionary demonstrations. An examination of the stars that she was born under will help establish a clearer understanding of the importance of the fixed stars in genethliacal astrology.

Patty Hearst was born at 18:01 hours on February 20, 1954 in San Francisco, California. The most prominent of the stars she was born under is the red alpha-Tauri (α-Tau). This first magnitude star was culminating at the time of her birth. In ancient times it was called the Torch, but later it was named Aldebaran and is the bright star in the Hyades marking the eye of the Bull. Aldebaran is astrologically as the nature of the planet Mars.

The Hyades in general and Aldebaran in particular have been the topic of poets such as Spencer (1552-1599), writing in "Faerie Queen," and ancient commentators of science, such as Pliny (23-79), writing in *Historia Naturalis* warned of its dire consequences. Firmicus Maternus, in his fourth century Latin treatise *Mathesis Libri VIII*, wrote that the red Aldebaran presages individuals who are "restless and riotous, always stirring up popular dissent and revolution." The star is said to inflame the minds of people with furious quarrels and to be the enemy of quiet and peace, madly desiring civil and domestic wars. But Aldebaran is also one of the most eminently fortunate stars in the heavens, portending riches and honor. Its natives will always be under stress and in a constant state of anxiety. However, they are always a success in pecuniary matters.

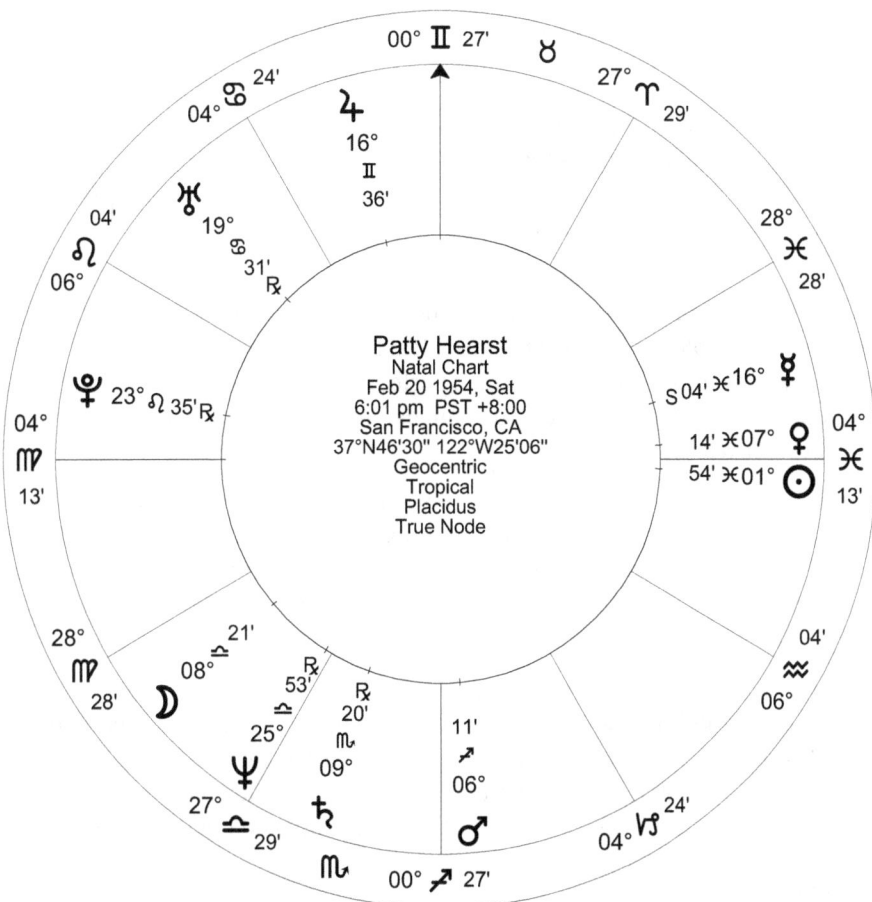

At the moment of Patty Hearst's birth Mars was on the Nadir in an exact opposition with the star Aldebaran. More than 1,500 years ago Maternus wrote that Aldebaran "when aspected by a malefic portends a sudden and unexpected involvement in riots and sedition resulting in being unjustly condemned by the people!"

Patty Hearst's rising star is gamma-Pyxis (γ-Pyx). This star, named Alphart, is the western star of the two located near the tip of the mast of the ship Argo. The constellation is indicative of those with sagacity. (Remember that the prow of the Argo was constructed of a piece of the speaking oak of Dodona so the ship would be endowed with the powers of warning and guiding its crew.)

Like Aldebaran, Alphart is also of the nature of Mars. Rising, it indicates individual of high principles and integrity, and those that are not easily swayed by the sophistries of the moment. Such a person has a strong will, and a propensity to stand against all who would attempt to subjugate them. While such a native is willing to bend a bit for the sake of reasonable compromise, they are not willing to sacrifice principle for expediency. Indeed, if pressed too hard they are apt to reel and their rash actions can lead to their downfall.

The Hearst picture is complete with the Martian star kappa-Aquari (κ-Agr) setting on the western horizon. This star, named Situla, is situated in the water that flows from the hand of the Water Bearer. Situla puts the planet Mars squarely on each of the angles of Hearst's chart, and most certainly this is representative of the nature of Tanya.

The star Situla is associated with natives who are bold and clever. They tend to be dissatisfied quite easily, however, and may wander from one interest to another. It is said that those with Situla setting never complete what they propose. But this is really a misnomer. What is operative here is a tendency to be impatient with less than perfection, even with one's self. The Martian impetuosity is at the root of all these observed changes. Situla is also the star of those who perform their most important work in secret.

It is highly unlikely that this short account of the stars under which Patty Hearst was born is an entirely accurate characterization of either her personality or her life, but the descriptions of these stars are taken from ancient manuscripts such as that of Firmicus Maternus. So accurately do parts of these descriptions fit the Hearst saga that it is hard to believe they were written almost two thousand years ago.

Most of what has been said above about Gerald Ford and Patty Hearst can, of course, be delineated from their charts without the aid of the fixed stars. but these latter elements supplement the basic chart and go far toward aiding in a comprehension of all that is contained in it. Many times classical astrologers put more emphasis on the fixed stars than they did on the planets. The next chapter will relate more of the applications of the fixed stars and constellations.

Review Questions

1. Planets on the angles or in aspect to the angles are also in aspect to the stars that are rising, setting, or culminating. For example, a planet trine the horoscope (Ascendant) is also trine any fixed star that is rising at that time. Comment on the chart of Gerald Ford.

The next two questions assume that the reader has a copy of *Classical Scientific Astrology* (or the reader's own chart can be used).

2. Using the formula in Appendix C (or the table or risings, settings, and culminations included in that appendix), determine the stars of the native in Chapter II of *Classical Scientific Astrology*. Alternately, determine your own stars.

3. Are there any differences between the lunar and solar chart regarding the stars the native of question two is born under? Explain.

4. What further information can be gleaned about the native of question two from the stars she is born under?

Chapter VI

Judicial Astrology

Definition and Requirements

The modern science of genetics stems from the work of Mendel in the nineteenth century, and a systematic study of the effects of the environment on the nature and conduct of man is a product of the twentieth century. But two thousand years ago it was recognized that both of these factors, environment and heredity, must be taken into account prior to the delineation of a genethliacal chart. The problem of the prince and the pauper, both born at the same instant in the same place, was not a problem at all to the classicists. The two identical charts would be interpreted differently according to the variances implied by the stations of life of the two natives. Even with identical twins there are subtleties in environment to be accounted for in interpreting their separate charts. In ancient times, for example, the one who emerged from the womb second was designated by law and custom to be discriminated against in favor of the first born. The effects of this discrimination on both of the natives was taken into account by the classical astrologers.

The above examples pertain to the factors of heredity and environment peculiar to the individual native. But of equal importance in assessing a genethliacal chart are the environmental and heredity considerations common to large groups of man. Custom and mores vary from one society to another, and the manner in which an individual will react to a given stimulus is conditioned in large part by these same customs and mores. In addition, the political and economic environment within which an individual lives impacts on all facets of his life. A chart indicating multiple marriages would be interpreted differently in Spain and the United States (and perhaps differently again in the several states). In the latter instance divorce would be suggested, while in Spain the death of the wife or concubinage would be more indicated. Mars prominent in a chart together with certain of the fixed stars and constellations (Bellatrix and Orion, for example) indicate military honors, but only if it can be foreseen that the native's country will go to war. Otherwise such a configuration is more likely to presage leadership in politics or business. Finally, the procurement of wealth in a land plagued by famine can mean merely getting enough to eat, while in a monarchy or dictatorship such wealth also implies the attainment of political influence. The list is endless. But, as all these examples show, the delineations of a genethliacal chart independent of the heredity and environmental background of the native is an exercise in futility.

Judicial astrology is concerned with those events and conditions that have widespread consequences influencing the lives of whole populations. Today judicial astrology is called mundane, or political, astrology and is more limited in scope, studying merely those events affecting the destinies of nations or other groups of people. Ancient judicial astrology had for its purpose the formation of a data base for use in the delineation of genethliacal charts. Modern mundane astrology is its own purpose, rarely impacting on individual charts, even those of the heads of state. This text uses the wider application, which includes modern mundane astrology as a subset. The classicists considered judicial astrology to be of greater validity than genethliacal, its predictions being based on somewhat firmer scientific foundation and more likely to be correct.

The scientific basis for judicial astrology was the classicists understanding of the mechanism of heredity. In terms of modern science this understanding was naive and superficial. But, stripping away the prejudices inherent in the parochialism of the era, there was a grain of truth in what could be readily explained by the then current science. As in *Classical Scientific Astrology*, the interest lies not with the truth of the ancient science per se, but rather with its truth pro forma. As far as classical astrology is concerned, if the positions of heavenly bodies can be correctly correlated with events on Earth, the explanation of why these correlations are correct is not relevant.

Astrological Geography

In *Classical Scientific Astrology* the relationships between the "winds" and the Aristotlean elements was discussed. As these elements were the basic constituents of everything in nature it is legitimate to ask, "What is the relationship between the winds, and the corresponding elements, and the observable differences between peoples of the different regions of Earth?" The answer was suggested at least four centuries before Ptolemy. Ephorus of Cymein Aeolis (d. 340 B.C.), a pupil of Iscrates, asserted that the winds blowing from the quarters of the Earth defined the peoples in these places. In his treatise, "On Europe," he wrote:

> "...if we divide the regions of the heavens and of the earth into four parts, the Indians will occupy that part from which the Apeliotes blows, the Ethiopians the part from which the Notus blows, the Celts the part on the west, and the Sythians the part from which the north wind blows...."

This suggests that the Earth can be quartered and that the nature of the people in each of the quarters can be related to the triplicities as discussed in Chapter III of *Classical Scientific Astrology*.

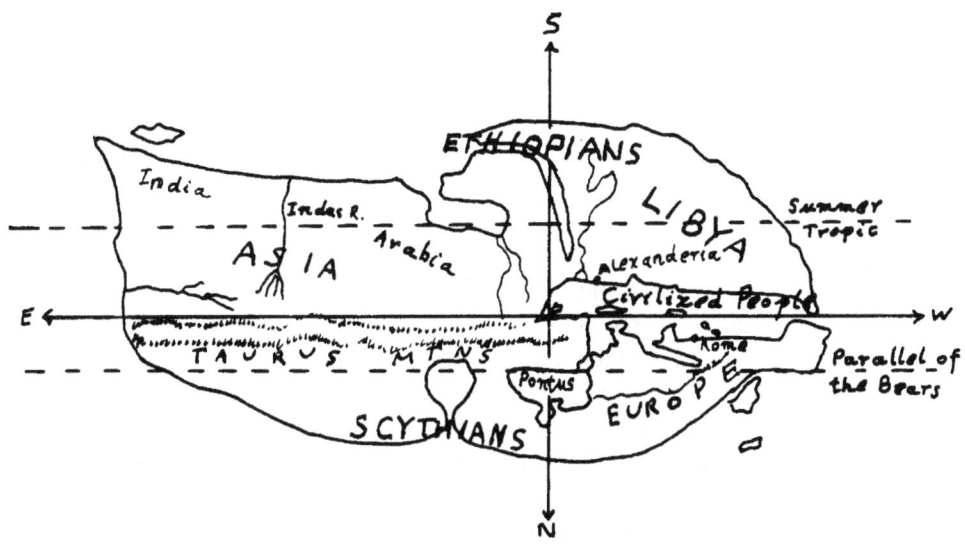

Figure 2. The Inhabited World According to Ptolemy

In terms of the geography prevalent at the time, this quartering was perfectly natural. It was recognized that the Earth was a sphere, but the populated portion of the Earth was assumed to be a trapezoidal surface on the sphere. Furthermore, the center of this populated Earth was the center of civilization: the Eastern end of the Inner Sea (Mediterranean Sea). The east-west line of the quartering runs from the Straits of Hercules (Gibraltar) to the Gulf of Issus, and along the mountains to the east (Taurus Mountains). The north-south line runs through the Arabian Gulf (Red Sea), the Aegean Sea, the Pontus (Black Sea), and Lake Maeotis.

Figure 2 depicts this quartering as imagined by Ptolemy. Note that the orientation of the map puts the south towards the top of the page so that the east is to the left. Such an orientation was the norm well into the Middle Ages. A comparison with modern maps will indicate some obvious errors in geography. The Taurus Mountains, for example, do not form a straight east-west line through Asia, and if the eastern point of the Mediterranean locates the north-south line, the Red Sea is crossed just south of the Gulf of Aquaba. Also, the inhabited world ends with India—the great Chinese civilization was unknown to the ancient Greeks. But the intent is clear and is quite logical in terms of the then current knowledge of geography.

The fiery trigon of Aries, Leo, and Sagittarius is indicative of the northwestern portion of the world and is governed by Jupiter and Mars. The southeastern portion of the world is given to Taurus, Virgo, and Capricorn governed by Venus and Saturn; Gemini, Libra, and Aquarius are northeastern and governed by Saturn and Jupiter; and the southwestern part of the world is given to Cancer, Scorpio, and Pisces, with the governors being Mars and Venus. In this scheme Jupiter denotes the quality of the north wind, Mars the west wind, Venus the south wind, and Saturn the east wind. The natures of these planets correspond to the natures of the various winds, and therefore to the general natures of the people that inhabit these portions of the world. But to account for the specific variations within each of these areas the natures, rulers, and triplicities must be utilized.

The demarcation of the civilized world in classical times is the summer tropic (Tropic of Cancer) on the south and the parallel of the Bears on the north (the latitude above which the constellation of the Bears never sets). South of the summer tropic it was postulated that the "excess of the Sun" made the people contracted in form, shrunken in stature, and sanguine in nature. North of the parallel of the Bears, on the other hand, the people are "tall and well nourished, and cold by nature." In these latitudes in ancient times the inhabitants were all considered to be savages: the Ethiopians in the south and the Sythians in the north. But between these two latitudes the people were (according to Ptolemy) all of moderate stature, equable of nature, and civilized. The eastern of those within the middle latitudes are of the nature of the Sun (masculine, with a vigorous personality), while those people in the western portion are of the nature of the Moon (feminine, softer of soul and a tendency to be secretive). Again, those that live in the southern section of the middle latitudes are said by Ptolemy to have a sagacious personality, and to be inquisitive and adept at mathematics (and astrology). These latter traits are correlative with the zenith being close to the zodiac in these areas.

To account for the differences between the people within a given quadrant, these areas were further subdivided by Ptolemy so that each subdivision was correlated to one of the signs governing that quadrant. For example, in the northwestern quadrant that part consisting of Britain, Transalpine Gaul, Germany, and Bastarnia were said to be in close familiarity with Aries and Mars. Italy, Apulia, Sicily, and Cisalpine Gaul have their correlative indications with Leo and the Sun, while Tyrrhenia, Celtica, and Spain were said to be subject to Sagittarius and Jupiter. Note that the mention of various "countries" in fixing the correlations between the sign and the various subdivisions of the populated world have nothing whatever to do with present day political boundaries. Neither was Ptolemy referring to the national boundaries of his day. The mention of these countries is merely a device to locate that portion of the world correlative with the natures of the distinct signs and planets. Considering the state of the knowledge of geography in those days it is instructive to ask what Ptolemy was attempting to accomplish. Modern astrologers blindly set down what was written without any consideration of the implications toward either the modern or the ancient scientific knowledge.

An analysis of *Tetrabiblos* (ii:3) in light of modern geography indicates a quite logical division of the world in terms of the then accepted data base: the science and philosophy of Aristotle. That portion of the globe north of 46N30 and south of the tropic of Cancer correlates with Mars and Saturn. This is consistent with the observations of the natures of the peoples in those areas at that time. In the middle latitudes those portions closest to the center of the populated world were considered most familiar with the Sun, Moon, and Mercury, while those farther out—the extremities of these middle latitudes—were considered of the nature of the beneficent planets Jupiter and Venus. The extremities (north, south, east, and west) are therefore correlative to the primary governors of the quadrants, while those portions near the center of the populated world are subject to the Sun, Moon, and Mercury. In the sixteenth century Hieronymus Cardanus published a commentary in which he asserted that Mercury was considered correlative to the middle portion of the world because the inhabitants of the central regions were more given to the arts and sciences and addicted to commerce, and because Mercury's nature lies midway between those of the other planets. Charles E.O. Carter, in *An Introduction to Political Astrology*, asserts that this scheme of Ptolemy's ". . . was decidedly pre-scientific" and should therefore be discarded. But from the foregoing it can be seen that nothing can be farther from the truth. Ptolemy's quartering was both scientific (in terms of the then known data base) and consistent with observation!

Figure 3 presents the quartering of the world according to Ptolemy in terms of modern geography. Mars, Jupiter, the Sun, and the Moon are relegated to the western portion of the world, while Saturn, Venus, and Mercury have the eastern portion. Note the symmetry that was so important in Greek thinking. The shaded portion in the center designates an arrangement similar to this basic quartering, but of an opposite nature.

The center of the populated world (according to Ptolemy) is designated by 35N00 latitude and 35E00 longitude (Figure 4). This places the east-west line a bit south of the Straits of Gibraltar, but preserves some of the other statements of Ptolemy. 35N30 might be a better value for the latitude, but for the purposes of this book the map of Figure 4 will be considered operative.

The north portion of the center of the world between 35N00 and 46N30 and from 06E30 to 19E00 is correlative of the Sun and Leo. From 45E00 to 64E00 in this northern portion of the middle latitudes Mercury and Gemini rule. In the southern portions between these latitudes the rulers are the Moon and Cancer, and Mercury and Virgo, respectively.

That portion of the world bounded on the west by longitude 19E00, on the east by 45E00, on the north by latitude 46N30, and on the south by the Tropic of Cancer are designated the central regions of the populated world. It was here that civilization began. According to the ancient science the central regions are a microcosm of the rest of the world. These regions must therefore be correlative of all the signs and planets, just as are those portions of the world outside this region. The difference lies in that the correlations are opposite those outside the center of the world, the northwestern portion of the central region having a familiarity with the signs and planets of the southeastern world outside the central region. The same holds true for the other quarters of course, so that the northeast central region has the same rulers as the southwestern outer regions, etc. The various subdivisions of the central regions and their familiarity with the signs and planets are shown in Figure 4. Hence Judaea is ruled by Aries and Mars, and Ptolemy mentions the resultant boldness of the inhabitants of that area that is evident even today. (The map in Figure 4 lacks the symmetry of that in Figure 3, and the east-west line is south of where Ptolemy places it in *Tetrabiblos*. The boundaries given in Figure 4 are suggestive only. The main thrust is believed correct, but the details are subject to change with further research.)

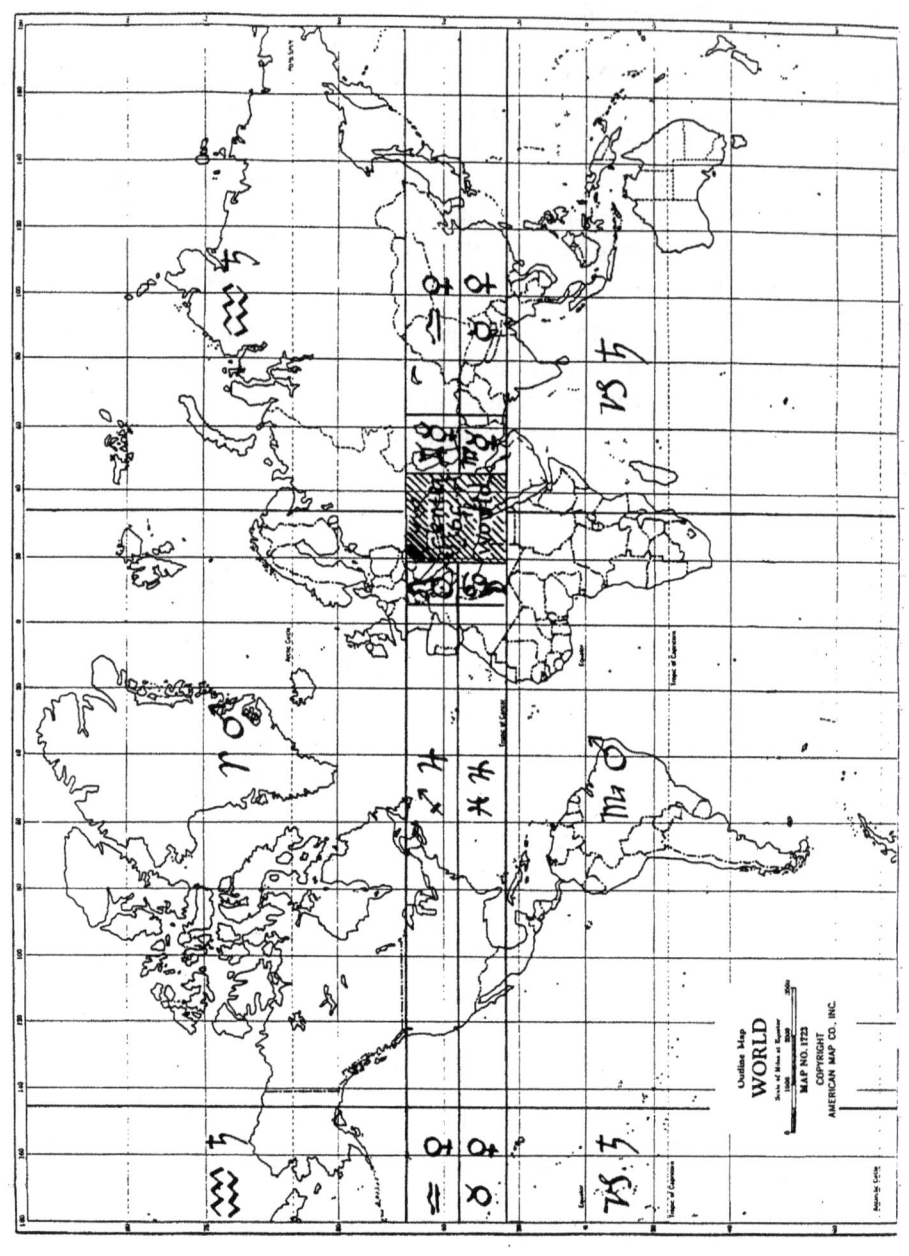

Figure 3. Ptolemy's Quartering of the World

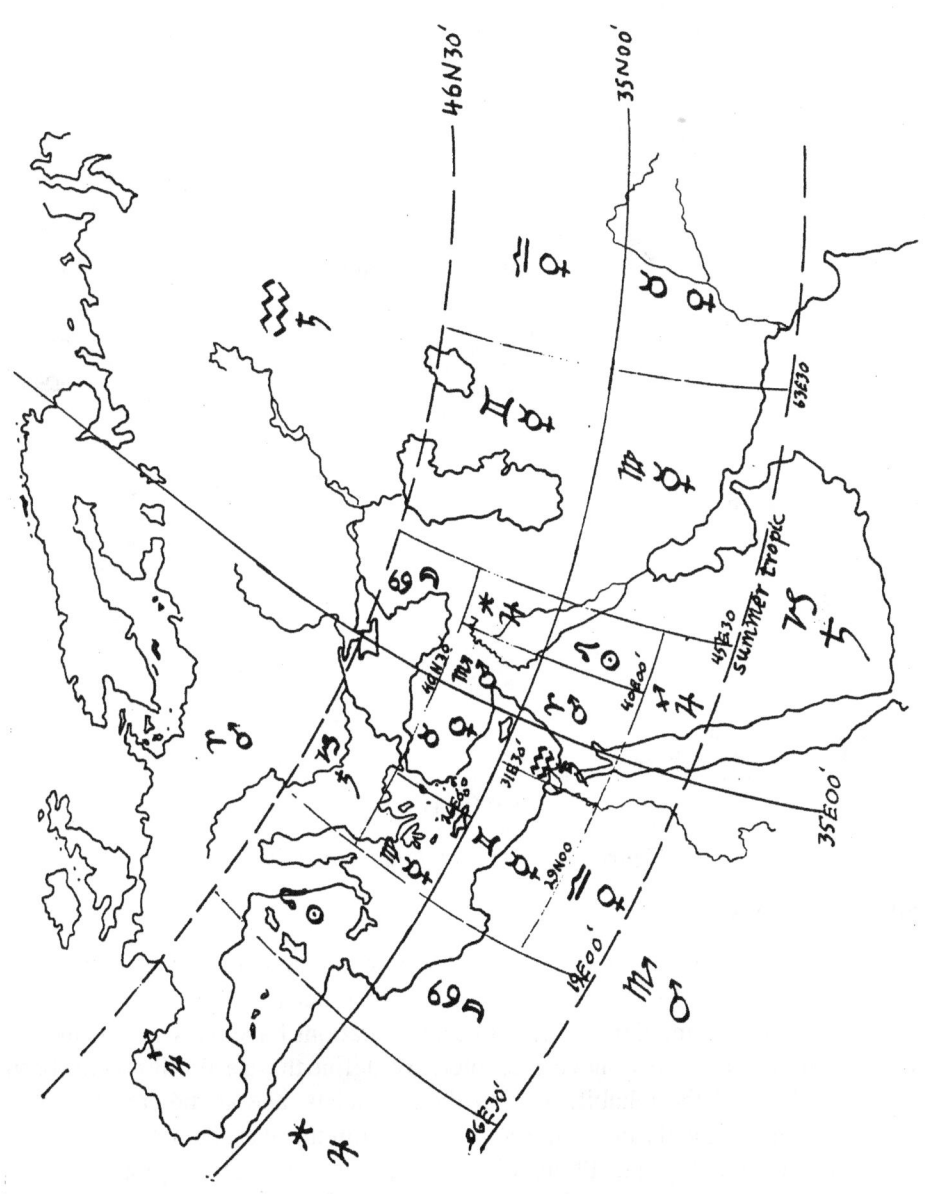

The Astrological Relationships in the Center of the World

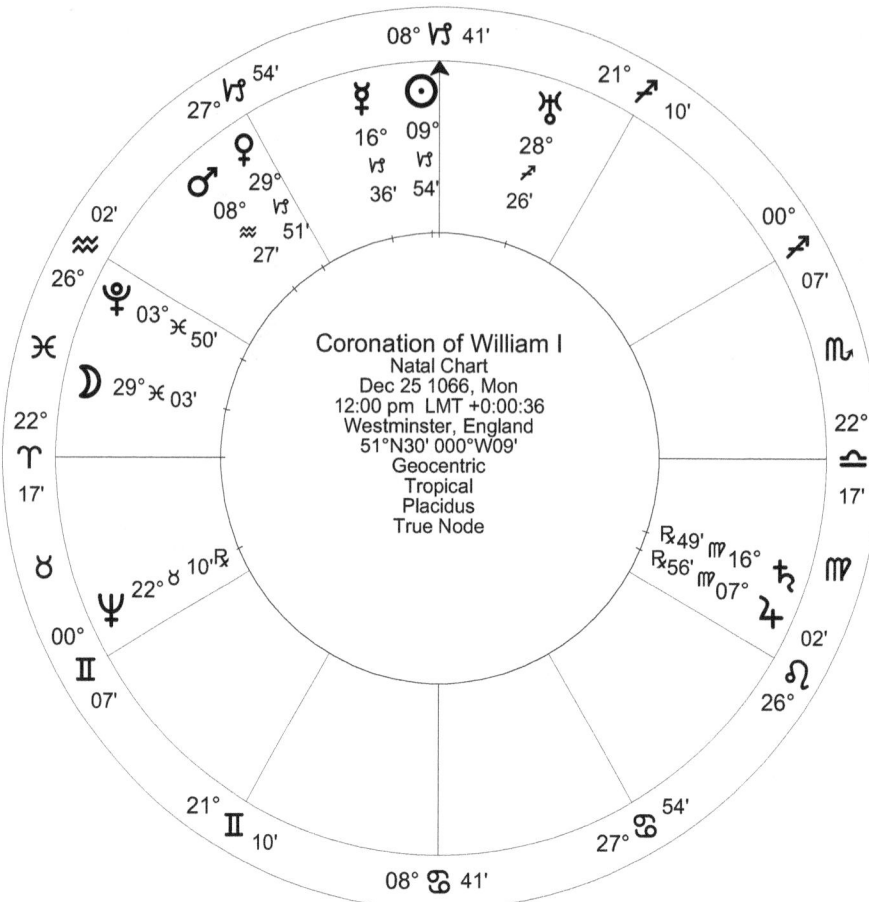

Figure 5. Chart of the English Nation

National Horoscopes

Ptolemy devised a scheme whereby various wide portions of the world were given a familiarity with the signs and planets. Modern astrologers, on the other hand, have misinterpreted this scheme and have either assumed that twentieth century nations have the same affinities as does the Ptolemaic region as a whole, or have attempted to redefine these affinities on the basis of the supposed personalities of the inhabitants. But the classicists take as the correlative sign and planet for a nation the sign and ruler of the Midheaven from the chart depicting the inception of the country as a political entity. The Ptolemaic sign and planet ruling the region in which the nation is located can still be expected to have a prominent influence in its chart, however.

Consider the chart for the coronation of William the Bastard depicting the inception of the English nation (Figure 5). Capricorn is on the Midheaven and this point is ruled (in the classical

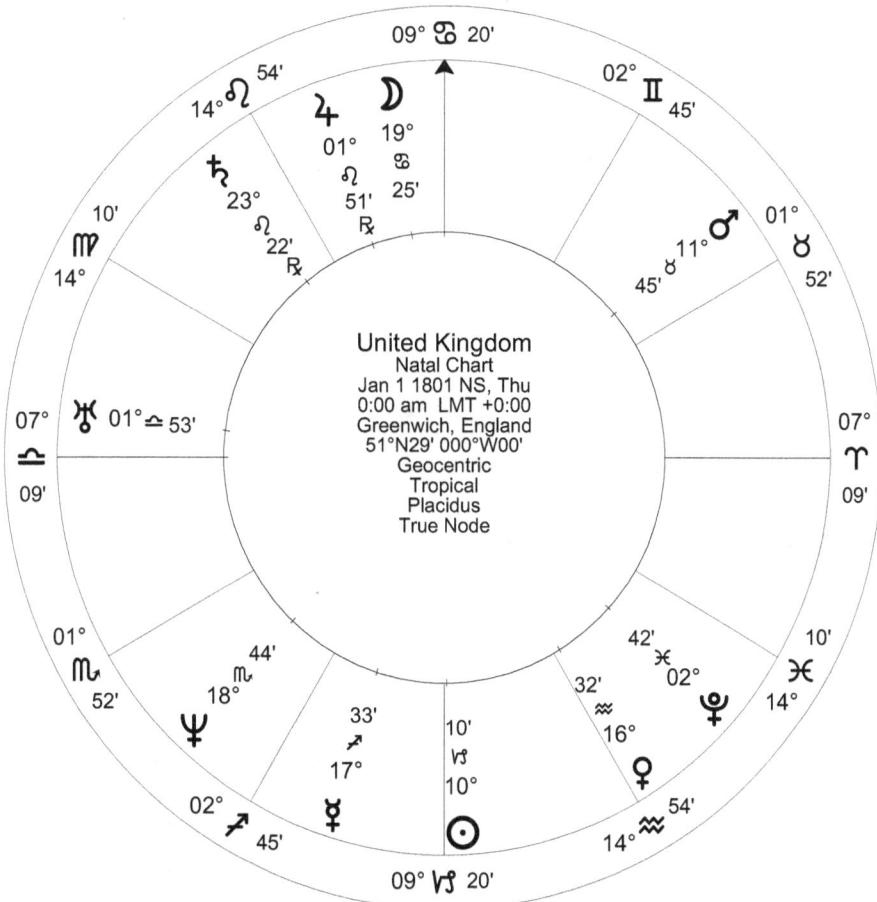

Figure 6. Chart of the United Kingdom

sense; see Appendix B) by the planet Saturn. The planet and sign correlative to England are, therefore, Saturn and Capricorn. But Aries is on the horoscope, and the horoscopic point is ruled by Mars. Furthermore, Capricorn is the exaltation of Mars. So while the English nation as a political entity has an affinity with Saturn and Capricorn, the correlatives of the Ptolemaic region of which England is a part (i.e., Mars and Aries) are very prominent in the chart.

The modern nation of the United Kingdom dates from the moment of the coming into force of the Act for the Union of Great Britain and Ireland in 1801. The chart of this event is shown in Figure 6. From this chart it is seen that the United Kingdom is ruled by Jupiter and Cancer. But note that Aries is on an angle and that Venus (the ruler of the horoscope) is quartile dexter separating to Mars. Furthermore, Mars will shortly be applying quartile dexter to Saturn, the ruler of the chart of the English nation. Here again the nation has as its correlatives that planet and sign taken from the nativity of the political entity, but the Ptolemaic sign and planet for the region as a whole play an important part in the chart.

In general it is impossible to draw up a national chart with any certainty of its validity. But there are some exceptions. The State of Israel dates from 1600 hours local summer time, May 14 at Tel Aviv; the Japanese constitution, fostered by the United States occupation forces, went into effect at 1730 hours local standard time, March 5, 1946 at Tokyo, according to *Japan's Imperial Conspiracy* by Gergamini; and the inception of the modern nation of Burma was chosen by its astrologers to be 0420 local standard time, January 4, 1948 at Rangoon.

United States

As an example in bringing the Ptolemaic theory up to date, consider the central latitude for the quartering of the world. This line cuts the United States just about in half at the Mason-Dixon Line (Figure 3). While the revolution against England started in the north it was the southern states, especially Virginia, that provided the intellectual momentum for the foundation of the republic. Ptolemy states, in *Tetrabiblos*, of those people correlative of Jupiter and Pisces: "...they are free and simple in their character, and they are worshipers of Jupiter as Ammon (a one and only God)...." And of those people who are correlative of Jupiter and Sagittarius: "...they have a propensity towards independence and simplicity...." Such a description certainly depicts the inhabitants of the New World in 1776, and suggests a possibility of constructing an inception chart of the United States.

From Figure 3 it is seen that the southern portion of the Untied States is ruled by Jupiter and Pisces, while the northern portion is ruled by Jupiter and Sagittarius. This suggests that Jupiter should play a prominent role in the chart of the United States, and that either Sagittarius or Pisces should be on the Midheaven (with a preference to Pisces with its relationship to the southern section of the nation).

Casting a chart for July 4, 1776 at Philadelphia with Jupiter rising places Pisces on the Midheaven. This places the Midheaven in the term of Jupiter (see Appendix B), as well as being in the lunar house of Jupiter in a nighttime chart (the time of the chart is 3:57 a.m.). (For those who have not read *Classical Scientific Astrology*, the terms are extremely important irregular divisions of the signs. That the Midheaven is in the term of Jupiter clearly intensifies the Jupiter relationship to this chart of the United States.) Such a chart is also consistent with the historical adage that the Declaration of Independence was finally completed in the early morning hours of July 4, 1776! This chart (Figure 7) is at variance with most modern charts that have Gemini rising on "psychological" grounds. But as was pointed out in Chapter II of *Classical Scientific Astrology*, it is the Midheaven, not the horoscope, that is most important in a national chart. The principles of classical astrology point most assuredly toward such a chart as the one depicted in Figure 7.

National charts such as those depicted in the last three figures must be tempered with the charts of the rulers of the country in question. Surely events that affect the rulers will also be of importance

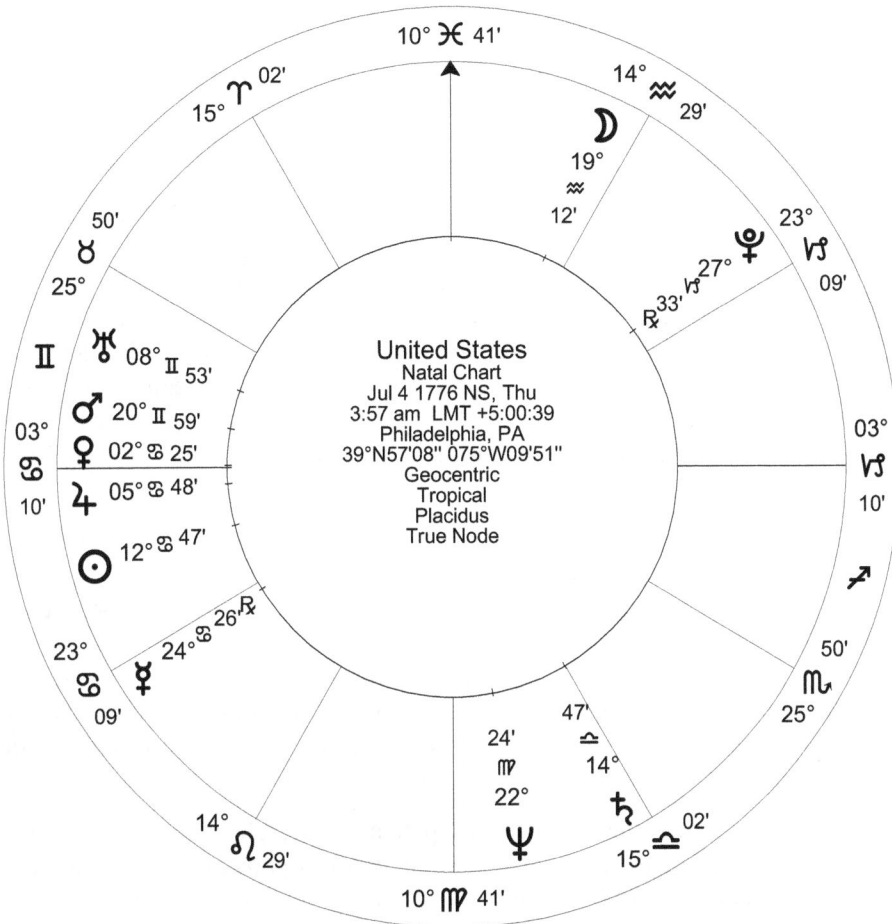

Figure 7. Chart of the United States of America

to the nation as a whole. In the same vein the inception of a new government, such as the inauguration of a president or the coronation of a new king, is also indicative of a chart that will affect the nation for the duration of that government. But the national chart is correlative of the more long-lasting events that affect the polity as a whole. Where possible it should be used in any delineation concerning judicial astrology.

At a more local level the chart of a city or a state will presage events that concern the inhabitants of that city or state. Here again definitive charts are almost impossible to come by. An exception is the State of California (1000 hours LMT, July 4, 1846, at Monterey—this time represents the surrender of Monterey to the Untied States; that is, the initial raising of the Stars and Stripes over California) and some of the mission cities of California (e.g., Presido of San Francisco at 5:28 a.m., July 28, 1776 and Mission San Diego at 5:21 a.m., May 17, 1769). Such documentation is seldom available, although some modern astrologers give lists of the sign affinity of the various cities (see, for example, *An Introduction to Political Astrology* by C.E.O. Carter). But again their sources are questionable.

The Astrology of Eclipses

By far the most important celestial phenomena in judicial astrology is an eclipse of the Sun or the Moon. These events not only provide indications concerning the future of nations or of other large groups of people, but can also presage significant turning points in the life of an individual. Today the eclipse has lost much of its power as a prognostic tool. This is due to the naive and often erroneous manner in which it is utilized in modern astrology. Today it is often thought that the time of conjunction or opposition in longitude marks the time of the eclipse and that eclipses, according to C.E.O. Carter, "...are zodiacally identical the world over...only the house positions alter according to the latitude and longitude of the places for which the various horoscopes are erected."

Nothing could be further from the truth. As always in classical astrology we start with the observed phenomena. If the eclipse is not visible (at least potentially) at the place in question it will have absolutely no effect on that location or on the people therein. In addition, the number of civil hours of the duration of the eclipse varies with location (remember that a civil hour is one-twelfth the time from sunrise to sunset daytime hours or one-twelfth the time from sunset to sunrise nighttime hours), and in the case of a solar eclipse the degree of obscuration is also different in the various parts of the world. Both the duration of the eclipse and the degree of obscuration have a marked effect on the delineation of an eclipse.

The concept that a phenomenon has to be at least potentially visible to have a viable effect is one that troubles most modern astrologers. The conjunction or opposition affects all charts, even those constructed in the daytime when these events are not visible. And, continue our modern critics, since when has a cloud cover altered the delineation of a chart? But the modern astrologer misses the point. An eclipse is not merely a conjunction or an opposition. It is an event in which the light of the Sun or the Moon is darkened. The eclipse lasts for a short period of time and does not affect all portions of the globe. Even a location that is in daylight at the time of a solar eclipse will not necessarily be affected—at that location the light of the Sun is not darkened and if the Sun is not darkened there is no eclipse as far as the people at that location are concerned! Of course, cloud cover never affects a chart. If the eclipse or any other celestial phenomenon occurs at the location in question it will be effective even if it can't be seen because of cloud cover.

To expand on this, a country (or an individual) will be affected by an eclipse if two conditions are met. First, of course, is the visibility criteria discussed above. That is, did an eclipse actually occur at the location in question (nation's capital or city wherein the native resides)? In addition, to be effective, the eclipse must be related to the country or individual in at least one of four ways: a) by the eclipse being located in a sign of the same triplicity that rules the nation (the sign on the

Midheaven determines the ruling triplicity for both the nation and the ruler of the nation; the sign on the Ascendant determines the ruling triplicity for other individuals); b) by the eclipse being located in the same sign as is the nation's or individual's horoscope; c) by the eclipse being in aspect with one or more of the luminaries in the nation's (individual's) chart; or d) the eclipse being in aspect at the time of the eclipse with the ruler of the nation's Midheaven or with the ruler of a non-reigning individual's horoscope. In this regard those familiar with modern astrology should be particularly cognizant of the classical method of determining rulership of a point as explained in Chapter IV of *Classical Scientific Astrology*. This method allots five points to the planet whose domicile the point in question lies, four points for the exaltation, three points for the term, two points for the trine, and one point for the decan. However, in the case of a national chart five points are awarded for the exaltation and only four points for the domicile. The planets associated with the terms, trines, and decans are fully described in *Classical Scientific Astrology*. A brief account can be found in Appendix B of this volume.

Clues to the nature of the event presaged by the eclipse come from the constellation of the star culminating at that time, from the constellation of the most brilliant star in stellar aspect to the angle on the chart preceding the eclipse and from the planet(s) that rules the eclipse and the angle preceding the eclipse. If the same planet does not rule both the eclipse and the angle preceding the eclipse the preference is given to the lord of the eclipse. If more than one planet is found to be the ruler a judgment must be made as to which is stronger in terms of the eclipse. For example, that planet that is closer to an angle, that is rising or has a greater acceleration, or is in the same sect as the eclipse would be considered the stronger.

The extent and intensity of the event presaged by the eclipse is determined from the ruling planet and from the degree of obscuration of the eclipse. The greater the degree of obscuration (amount of the Sun's or Moon's disk shadowed), the greater the intensity of the event. That is, the more damaging the earthquake or flood, the longer the drought, the greater the economic recovery, and the like. Also, if the ruler of the eclipse is occidental to a solar eclipse or oriental to a lunar eclipse it is said that the event will affect only a minority of the people of the nation in question. If the ruler is in opposition to the eclipse then about one-half of the people will be affected. Finally, if the ruler is oriental to a solar eclipse or is occidental to a lunar eclipse then the effects will be widespread indeed, involving the majority of the people of the nation.

If the planet ruling the eclipse is also the ruling planet of the nation in question, and is also a benefic planet, then the eclipse presages good for the nation and all the more so if the benefic planet is not overcome (that is, in aspect or besieged by a malefic) by any of the malefics. If the ruler of the eclipse is also the ruler of the country and is a malefic the indications are for a more neutral effect shading to good if overcome by the benefics. But if the ruler of the eclipse is not also the ruler of the nation the indications are bad; they are particularly destructive if a malefic planet rules the eclipse, does not rule the nation in question, and is not in aspect or besieged by the rulers of the country. If either of the luminaries in the chart of the nation or native are in the same locus (house) or locus directly opposite of that of the eclipse in its own chart then the native or nation will most certainly be affected by the eclipse. The effect is generally bad under such conditions and if the luminaries are in these loci and at the very degree of the eclipse the indications are very destructive indeed. What has been said of the luminaries also applies to the angles, especially the horoscope and Midheaven.

The quality of the event indicated by the eclipse is determined by an analysis of the fixed stars and planets we have been considering. Aspects to other planets must be regarded, as well as their locations (in sign, triplicity, term, etc.). The nature of the planets and fixed stars are of primary importance.

Saturn is correlative of destruction by cold. If the event concerns man the indications are for illness, exile, poverty, imprisonment, mourning, fears, and death (especially among the old). As concerns the weather the indications are for fearful cold, freezing, and destructive snowstorms and pestilential weather in general. Very high and very low tides may also be presaged with storms at sea and the pollution of rivers and streams. In ancient times the destruction of whole fleets of ships was envisioned. As regards agriculture in general, Saturn alone with no mitigating factors would be correlative to the loss of livestock through disease, a scarcity of fish, and a loss of crops (even famine), through disease, insects, floods, or hail.

Jupiter is correlative of an increase of that which is beneficial to man. If the event concerns man himself, Jupiter alone will presage fame and prosperity, abundance, peaceful existence, an increase in the necessities of life, body and spiritual health, and (in medieval times) benefits and gifts from rulers would be indicated. In other areas this planet presages good weather favorable for crops and an increase in the economic good for the nation or individual.

Mars is indicative of destruction through dryness. This planet is correlative with wars, civil faction, capture, enslavement, the wrath of leaders and sudden death arising from such causes. In ancient times Mars was said to cause violence, assaults, lawlessness, arson, murder, robbery, piracy, and a violent death especially in the prime of life. If the event indicated by the eclipse concerns the weather then it can be expected that there will occur hot weather with pestilential and withering winds. There will be a drying up of springs and the pollution of potable waters. In agriculture, droughts can be expected with the resulting loss of crops. Medieval astrologers also attributed plagues of locusts to a Martian eclipse.

If the ruling planet and applicable fixed stars of an eclipse are of the nature of Venus the effect will be much as was discussed under Jupiter. If the indicated event concerns man there will be fame, honor, happiness, abundance, a good marriage, and in general a good life. The weather will be good, but with generous showers that will enable crops to grow. The economy of the nation will prosper with an abundance of crops.

Mercury, of course, takes its nature from the planets and fixed stars that it is in aspect with. In relation to an eclipse this planet indicates events which concern religion and changes in custom and laws of a nation. Changeable weather is also presaged with hurricanes, thunderstorms, and earthquakes. If Mercury is setting, rivers are likely to run dry, while if the planet is rising the indications are for full-flowing streams and rivers.

The above summary of the events correlative to the natures of the five planets could have been gleaned from what has been written in our companion volume (or many other astrological texts). The effects of the luminaries themselves as rulers of an eclipse will be left as an exercise for the reader.

The timing of the events indicated by an eclipse is determined through an analysis of the length of time the phenomenon is observable at the locality in question. A solar eclipse will affect conditions in a given locality for a period of time equal to the number of years as there are civil hours that the eclipse is visible. For example, if a solar eclipse lasts for three civil hours then the effect of the eclipse can be expected to last for three years. Lunar eclipses last for a shorter period of time than do solar ones. The effect of a lunar eclipse therefore is as many months as is the duration of the eclipse in civil hours at the locality in question.

The events indicated by the eclipse will begin to occur sometime during the first four months after the phenomenon if the eclipse is on the horizon, during the second four months if the eclipse is on the Midheaven, and during the third four months if the eclipse is on the Descendant. If the eclipse is on the IC it will not affect the locality at all. Why?

The time of greatest intensity of the event is taken from the location and duration of the eclipse. If the eclipse is on the horizon the greatest intensity will occur during the first third of the time that the phenomena is effective. If the eclipse is on the Midheaven look to the second third of the period of effectiveness of the eclipse for the greatest intensity and if the eclipse is on the Descendant the greatest intensity will occur sometime during the final third of the period.

These are all general indications of the timing of the event. And they must be consistent! If the duration of the effectiveness of the eclipse is of a shorter time than the number of months to the start of the event then, of course, the event cannot occur. For example, a lunar eclipse of three civil hours duration located on the Midheaven will have no effect whatsoever on the locality in question, for the start of the event will not take place for at least four months but the eclipse is effective for only three months.

The actual trigger of the event (that phenomenon that will set in motion that which is indicated to happen), is a conjunction of the ruling planet during the time the beginning of the vent is expected to occur. If it so happens that the ruling planet does not come into conjunction during this period of time the event will not occur or if it does occur it will be greatly abated in its effect. If the ruling planet at the time of its conjunction is rising or stationary the event will be greatly intensified. If the planet is setting or under the rays of the Sun at the time of the conjunction the event will be abated in its effect.

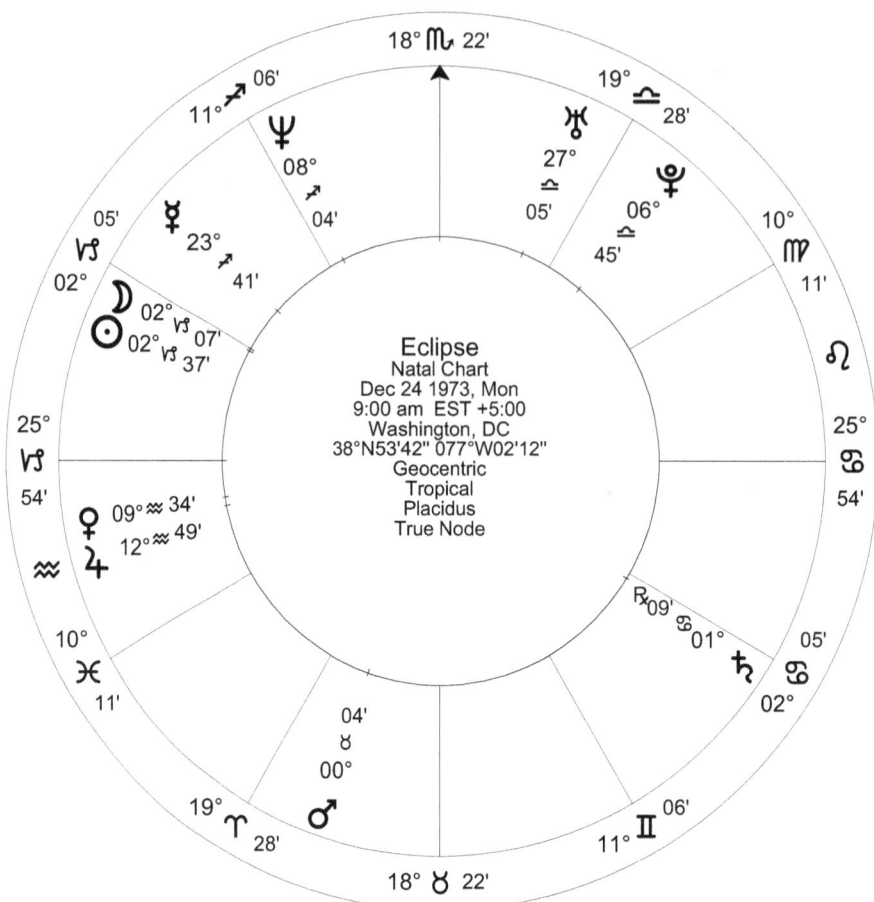

Figure 8. Chart of the Annual Eclipse of December 24, 1973

The Eclipse of December 24, 1973

An example will aid in firming the concepts of the theory of eclipses in classical astrology.

On December 24, 1973 there occurred an annular eclipse of the Sun. The approximate time of the middle of the eclipse at Washington DC is 13:55 GMT (8:55 a.m. EST). The semi-duration is approximately seventy-five minutes. This latter time must be subtracted and added to the middle of the eclipse to give the times for the beginning and ending of the eclipse. The duration of the eclipse is therefore 150 minutes or two hours and thirty minutes. More precise calculations using the method given in Appendix A shows that the eclipse begins at 12h59m59s GMT and ends at 15h09m05s GMT. The middle of the eclipse is at 14h00m35s GMT. With these more precise calculations the duration of the eclipse is now 2h09m07s. The approximate values of these data are sufficient for most purpose, but the more precise values should be used where possible.

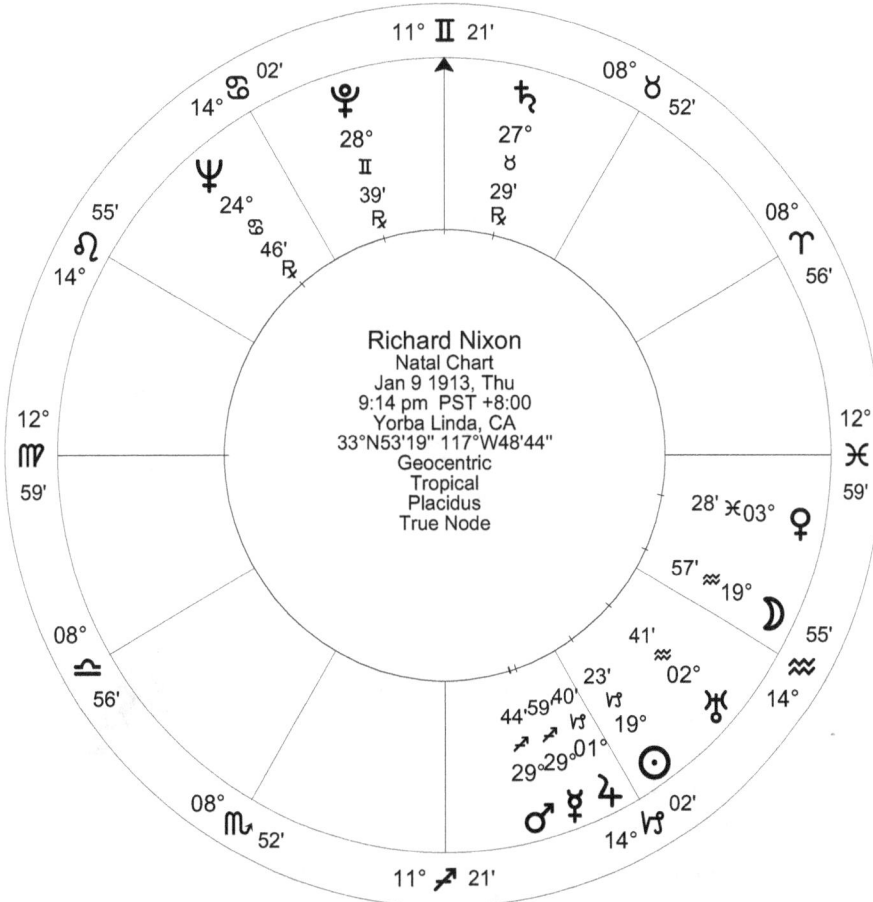

Figure 9. Chart of Richard M. Nixon

A chart is now erected for the middle of the eclipse at Washington, DC (see Figure 8). Note that the middle of the eclipse is not the instant of the conjunction of the Moon and Sun. Using the Sun's position (in a solar eclipse) note that the location of the eclipse is 2 Capricorn 28 and close to the horizon. The eclipse can be seen at the location of Washington, DC, however. In addition, Jupiter in the chart of the United States (the ruler of the nation) is opposed to the eclipse almost to the very degree. Hence this eclipse will most certainly affect the United States.

Now consider the chart of the nativity of Richard Nixon (Figure 9), who was then president of the United States. The Moon in Nixon's chart is in the first locus (house) opposite that of the eclipse. Also note that Nixon's horoscope is in the same triplicity (earth) as is the eclipse in Capricorn, and the ruler of Nixon's Midheaven, Mercury, is applying to a conjunction with the eclipse! And so both the United States as a nation and its president will be affected by the annual solar eclipse of December 24, 1973.

In the late fall of 1973 the United States was in a political turmoil over the Watergate scandal. In July of that year Mr. Butterfield had let drop that all of President Nixon's conversations had been taped. By December his political enemies were calling for his impeachment. The astrologers in the United States and abroad were publicly predicting that the president would weather the storm. But the astrologers reckoned without an analysis of the eclipse of December 24, 1973.

The rulers of 2 Capricorn 28 are Mars and Venus (see Appendix B). the ruler of the horoscope of the eclipse is Mars alone. So Mars is the ruler of the eclipse. The star Dabih (β-Cap) is rising and rules the eclipse, and the star Merez (β-Boo) is culminating. As both the location of the eclipse and the star rising at the time of the eclipse are in the constellation of Capricorn, political changes can be expected (see Chapter III). From the temper of the times anything, including revolution and/or assassination, would be quite possible. Mars, being the ruler of the eclipse, could certainly mean either of the above possibilities. But Mars is occidental to a solar eclipse and does not rule either the nation or its president. From the occidental placement we deduce that only a minority of the people will be affected. Hence revolution is out! But what of assassination? Mars is not the ruler of Nixon's chart, nor is this planet overcome by either Venus or Jupiter. But the Moon (and the Sun) is trine dexter separating Mars and Mars is sextile dexter applying Saturn. Seems to be a stand-off and assassination is still a possibility.

On December 24, 1973 a civil daytime hour at Washington, DC was 48m01s.8 (or 0.8 equinoctial hours, see Appendix A). Therefore the duration of the effect of the eclipse will be 2.7 years (or until November 1, 1976; that is, until the election of a new president!). As the eclipse is located close to the horizon at Washington, DC, the beginning of the event will be within the first four months (prior to April 24, 1976) and the greatest intensity within 10.8 months from the time of the eclipse (by November 15, 1976).

Mars rules the eclipse and on April 20, 1974 comes into conjunction with Saturn! On April 19, 1974 the investigating committee of the United States House of Representatives voted to demand of President Nixon a transcript of the tapes of his conversations vis-a-vis the Watergate coverup. Nixon received this demand the next day (April 20), setting in motion the events that were to lead to his resignation from office on August 9, 1974!

The conjunction of Mars and Saturn takes place in Nixon's tenth locus (house), and is almost in an exact opposition with Mercury (the ruler of his Midheaven) in Nixon's natal chart. This eclipse has to be detrimental to Mr. Nixon's career! But Mars and Saturn are setting at this time (of the conjunction) and the easy aspects with the Moon and Saturn at the time of the eclipse all indicate an abatement of the effects. Furthermore, the magnitude or obscuration of the eclipse is only 0.355. Assassination is out. Given the political realities at that time, only impeachment or resignation is a viable prediction. (A given astrological configuration may have many possible interpretations. Therefore, delineations after the fact such as we have just done are comparatively easy. But in this instance at least the results should have been straightforward, at least to the final result of the disgrace of President Nixon.)

The Syzygy

Eclipses are just a special case of a phenomenon that happens every month: the syzygy, or full and new Moons. Most of the time neither the Sun is obscured by the Moon at the time of the new Moon, nor is the Moon obscured by the Sun at the time of the full Moon. Nonetheless the syzygy is always an important astronomical event from the astrological point of view. Most charts erected to predict events in judicial astrology are timed to correspond to the instant of a syzygy. The most important of these is the new Moon of the year, which is defined as the syzygy that most nearly precedes the vernal equinox. (In earlier times the Moon of the year was restricted to the new Moon only. However, as the new Moon preceding the vernal equinox can be an entire lunar month earlier the procedure was changed.)

The chart of the Moon of the year erected for the location of a nation's capital is used to predict events pertaining to that nation for the coming year. The procedure is analogous to that described previously under eclipses. The ruling planet and the stars in stellar aspect, together with the constellations, will indicate the general tenor of the coming year. In addition, because the Moon of the Year is that syzygy that most nearly precedes the vernal equinox, the chart is of particular importance in delineating conditions for the three months of spring. The syzygy most nearly preceding the autumnal equinox and the solstices is used in like manner to predict the general conditions of the three months following each of these events.

The greater detail of a monthly prediction is made from a chart of the new or full Moon erected for the locality in question. If the syzygy of the preceding equinox or solstice was a new Moon then the next monthly chart should be cast for the new Moon. If the syzygy in question was a full Moon then the next monthly chart should be cast for a full Moon also. Some classical astrologers attempted to get even more precision in their predictions by casting charts for the quarter Moons, too. Such latter chart should be used with care, however, as the precision implied by their use is not generally obtainable (see Chapter II of *Classical Scientific Astrology*). Where it was necessary to attempt the precision of day-to-day prognostication, classical astrologers depended on the rising and setting of the most brilliant and powerful of the fixed stars and on the movement of the planets across the heavens. These stars, coupled with the general conditions implied by the syzygy charts, frequently auger daily conditions, especially as concerns the weather.

Astrological Meteorology

The most important application of astrology in classical times was near and long range weather prediction. Today, advances in the science of meteorology have left astrology far behind in this field except perhaps for the most long term of climatological predictions. Even here the new weather satellites will likely soon allow long range weather prognostication with a precision that astrologers can only dream of approaching. Such a condition is normal for an art such as astrology, because as man obtains greater knowledge of his universe and the forces therein he requires less dependence on mere random correlations for his predictive needs. Indeed, sometime these correlations themselves can be explained to such a depth that new laws of science are developed. But the ancient astrological method of weather prediction is quite interesting in itself. The classicists were astute observers and much of their weather astrology was remarkably accurate.

The Sun was observed at rising and setting to determine the weather for the following day and night. Solar aspects to the Moon and direct observations of the Moon and fixed stars gave the classicist weather predictions of longer extent. For example, when the Sun rose or set clear, unobscured, steady, and unclouded, the indications for the next twelve hours were for clear weather. But if the disk was variegated or reddish or sent out ruddy rays it was said to indicate heavy winds with the direction of the wind from the angles at which the rays pointed. If the Sun was dark or livid, or had halos and gave forth dusky rays, then storms and rain were portended.

The Moon is also observed. If it appeared thin and clear and has nothing around it the inference is for clear weather. If the Moon is thin and red, the indications are for winds, and if it is observed to be dark or pale and thick, it signifies storms and rain. If the Moon has a halo around it that is clear and gradually fading, it signifies fair weather. If the halos are yellowish and broken, the indications are for heavy winds; if thick and misty, then snowstorms; and if the halos are pale or dusky and broken, the prognostication is for storms with both wind and snow.

But such observations were already centuries old at the time of Ptolemy. Six hundred years previously Aristotle wrote in *Meteorologica*: "This information (halos) is therefore a sign of rain, while if it is broken it is a sign of wind, if it fades, of fine weather. For if it neither fades nor breaks, it is reasonable to regard it as a sign of rain." Aristotle here is talking not only of halos about the Moon, but also those about the fixed stars. Ptolemy adds to these observations, noting that if the clusters such as the Praesepe appear dim in a clear sky the significance is for heavy rains, but if these clusters are clear and constantly twinkle, the indication is for heavy wind. Also, the Asses on either side of the Praesepe indicate that a north wind will blow if it is the northern one that becomes obscured, and vice versa.

The above passages clearly demonstrate the emphasis that the classicists placed on observational astronomy. Such admonitions are noticeably lacking from modern astrologers, most of whom would be hard put to distinguish the planet Jupiter from the star Sirius in the night sky. The ancient astrologer/astronomers were excellent observers of the heavens (to the limit of their crude measuring instruments). But today these observations are neglected by all, including modern scientists. It would be well if modern science would investigate some of these observations, such as the one by Aristotle that correlates earthquakes with lunar eclipses.

Solar Ingresses

Modern astrologers also do not use the syzygy as the basis for their monthly or yearly predictions. Rather, in modern practice the solar ingress into the various signs is used instead. The chart is constructed for the instant of these ingresses. The English magazines *Modern Astrology* and *Old Moore's Almanac* rely on the ingresses, as do most American publications. Some of the better modern astrologers, such as C.E.O. Carter, do take the ingresses skeptically, but most do not. There is no real justification for the solar ingresses in classical astrology except perhaps the ingress into the vernal equinox that we shall meet presently.

Comets

From time immemorial comets have been important in astrology. Tycho Brahe foretold of Gustavus Adolphus through a consideration of the comet of 1577. The most famous of all comets, Halley's Comet, foretold the death of Agrippa in 11 B.C., the destruction of Jerusalem in 66, and the death of Philip II of France at Mantes on July 14, 1223. It also presaged the victory of Aetius and his Visigoths over Attila the Hun at the battle of Mauriacus in 451. The comet of 837 was one of the brightest of all times. Chinese records claim that its tail was a length of eighty or more degrees. Astrologically it portended the breakup of the Carolingian Empire and the beginnings of the modern nation of France. And, of course, the return of Halley's Comet in 1066 presaged the Norman invasion of England and the victory at the Battle of Hastings.

Ptolemy devotes two chapters of *Tetrabiblos* to discussion of the effects of comets and events that are generally considered to have little effect in judicial astrology (such as conjunctions of the Moon and Mercury) take on great significance when they occur in connection with comets. Comets have an astrological nature akin to that of Mercury and Mars. They traditionally bring wars, revolutions, famines, and the like. In classical times descriptions were given of their heads and tails. Ptolemy mentioned "beams," "trumpets," and "jars" as descriptors of the shape of various comets. Such descriptors are not used today because astrology is no longer connected with observational astronomy.

Like eclipses, a necessary condition for a comet to affect a given country is that it be visible in that country. The greater the brilliance (apparent magnitude) of the comet the greater the effect it will have. The chart is erected for the tie of the greatest brilliance of the comet or for the time it was initially observed or for the time of some other important astronomical event such as an eclipse that occurs in connection with the comet. The analysis continues as was described under the eclipse. But in this instance not only the constellation in which the comet is posited must be noted but also the constellation to which the tail points. The location (constellation) of the head of the comet determines the type of event that can be expected and the tail determines the nation, or portion of the earth, that will be affected by the event. The length of time the comet is visible is used to ascertain the duration of the event and the comet's position in reference to the Sun correlates with the beginning of the vent: the comet's appearance in the orient betokening rapidly approaching events, and in the occident that events that approach more slowly.

No better example of the result of ignoring these principles can be given than that of Kohoutek's Comet. On March 7, 1973 Dr. Lubos Kohoutek of the Hamburg Observatory discovered a tiny sixteenth magnitude comet. The first calculations predicted that the comet would come to within 0.1 A.U. of the Sun and based on some very iffy assumptions it was also predicted an apparent magnitude of as high as —6.5 was possible. Later much more conservative predictions were made. But the media latched onto the initial data and launched a Comet of the Century blitz. Astrologers from all over the world were consulted as to the probable meaning. Ephemerides of the comet were published in astrological magazines. Using the naive techniques of modern astrology the comet's future position was translated to the ecliptic and aspects by the planets were duly noted. The constellations and their meanings were completely neglected. There was no need to observe the comet—after all, astrologers never actually observe the planets. Astrology just wasn't done that way. And so predictions were then made and almost without exception all the astrologers foretold of horrendous happenings, including major earthquakes, World War III, and at least one of the end of the world. And then the big letdown. When the comet could finally be seen by the naked eye on January 15, 1974 it was but a magnitude 5 object (just barely visible). Needless to say, none of the major predictions of the astrologers came true. Not only was the apparition very dim, but it could only be seen south of latitude 35N00. That is, the comet could not affect any of the major industrial nations.

Major Conjunctions

Conjunctions of the superior planets are also used in judicial astrology. Of these, the most important is the conjunction of Jupiter and Saturn. The importance of this conjunction was first postulated in the Middle Ages by Abu Ma'shar, but legend has it that its use is much older. It is instructive to relate this legend.

By the beginning of the ninth century of the Christian era, the power of Islam had reached its high water mark. The previous century had seen armies turned back from Constantinople and the defeat of Abd El-Rahman at the Battle of Tours by Charles Martel in 732. And, as will be seen, in 809 there occurred an event of great astrological importance. Therefore, the Islamic kings turned to their astrologers for assurances concerning the future viability of their empire.

Tradition dies hard. Many centuries after the devastating deluge described in the Bible the Persians built a great building called the Sarawiya in the ancient city of Jay. Fearing another flood, the Persians cached manuscripts containing all of their scientific knowledge high up in the Sarawiya. Shortly after 800 this cache of documents was discovered and turned over to Abu Ma'shar, the court astrologer in Baghdad, for study. These manuscripts contained a remarkable theory that significant religious and political changes were indicated by a conjunction of Jupiter and Saturn. (This account is a slight idealization taken from *The Thousands of Abu Ma'Shar* by D. Pingress. Further background can also be found in *Astrological History of Masha'allah* by E.S. Kennedy and D. Pingree.)

Conjunctions of Jupiter and Saturn occur at intervals of about twenty years, and successive conjunctions tend to stay in the same astrological triplicity for more than two hundred years, after which they move into another triplicity in the order of fire, earth, air, and water. All conjunctions of Jupiter and Saturn are astrologically important, but a shift into a new triplicity indicates changes of a more sweeping nature such as the rise of a new nation. When, after more than eight hundred years, the conjunction shifts into the fire signs even more momentous events are heralded. With these newly found theories Abu Ma'shar proceeded to obey the admonitions of his masters to apply astrology for the glory of Islam.

Abu Ma'shar wrote an astrological history of the world based on conjunctions of Jupiter and Saturn: *Kitab Aluluf*. He correctly recognized the shift into the water triplicity in 571 as indicating the rise of Islam, but he relied too heavily on the inaccurate observations of the Persians. For example, he believed that successive conjunctions of Jupiter and Saturn remained in the same triplicity for two hundred forty years, and that the return to fire required nine hundred sixty years instead of an average of slightly more than eight hundred years. As a result there were some gross errors in many of his calculations.

More than a century later the great Islamic astrologer Masha'allah brought the world of Abu Ma'shar up to date in an astrological world history entitled *On Conjunctions, Religions, and Peoples*. In correcting the errors of his predecessor he was able to show, for example, that the shift into fire in 26 B.C. heralded the rise of a new force: the birth of Christ and the emergence of Christianity. Of the vernal equinox for the preceding conjunction of this event he wrote:

> "... when I looked at it I could find no planet stronger than the Sun, because it is in the Ascendant and it is the rod of its exaltation. It confers its counsel upon Saturn, and Saturn is in the ninth (locus), the position of prophecy, so that a prophet will be born; God will illuminate darkness with him, and give sight to the blind. And because Saturn, which receives the strength of the two luminaries from the Ascendant, is in Sagittarius, it indicates that his birth will be in the second conjunction from this conjunction, and because the aspect is trine, it indicates compassion and gentleness, and because of the locus of Saturn with respect to the locus of its house, it indicate what violence befalls him from his people."

In 7 B.C. occurred the great conjunction of Jupiter and Saturn that many say was the Star of Bethlehem. But of greater significance is the chart Masha'allah drew up for the vernal equinox of 3381 B.C. In this year there was a shift to the water triplicity and Masha'allah wrote that this chart indicates that the conjunction of Jupiter and Saturn following this one, in 3361 B.C., marked the biblical deluge. Recent excavations at Ur and modern scientific techniques have put a great flood at about 3350 B.C. But Masha'allah, using astrology and the theory of the conjunctions of Jupiter and Saturn, came to this same conclusion 1,000 years ago.

As noted above, at 2133 hours GMT on June 26, 26 B.C. there was a conjunction of Jupiter and Saturn at 3 Leo 01. This was a shift to the fire triplicity and indicated the ensuing influence of Christianity.

The next shift to fire was the conjunction at 4 Sagittarius 04 on October 4, 809 at 0057 hours GMT. A few years previously (800) Charlemagne had become emperor of the west. Rather than indicating a further greatness for Islam, this conjunction presaged the development of the great

European monarchies. From a more or less unified Roman rule, European society would not be transformed into a heterogenous nationalism.

The last shift into fire came at 0440 hours GMT on December 4, 1603 at 11 Sagittarius 49. This shift has indicated the sometimes violent change from oligarchy to a polity emphasizing the equality of all men. As can be seen, each of these shifts to the fire triplicity has portended far reaching changes affecting the very foundations of society on a worldwide basis.

These shifts to other triplicities have also had a major impact on mankind, but of a more limited nature. The shift into air in 1127 presaged the religious rebirth of Europe through the rise of monasticism. The shift into water in 1425 indicated the rise of England as a great world and naval power. And the shift to earth in 1802 has brought with it the great technological revolution that has put man on the Moon. The next shift will occur in November 2019 with the conjunction of Jupiter and Saturn in the air sign of Aquarius. What does this last shift indicate? One possibility is another return to God as answer to man' problems. Perhaps the readers of this book applying the principles given herein will be able to make a more definitive prediction.

Modern Confirmation of the Astrology of Major Conjunctions

It is about this time that a modern scientist will begin to laugh and to comment on how absurd this all is and that no intelligent human being in his right mind could possibly believe such garbage. Indeed, taken out of the context of the science of its day, it is absurd. What can be more ridiculous than considering earth, fire, air, and water as elements? And such superstitions as the sign of the zodiac having any influence on the people of this planet should be rejected out of hand. Furthermore, the locations or changes therefore of the conjunctions of Jupiter and Saturn indicating anything at all is just too stupid to comment on. But consider the following.

The Russians have long been interested in finding methods for predicting future events having a major impact on their society: famine, epidemics, depressions, etc. Professor A.L. Chizhevskiy (1892-1964) long ago recognized the possibility of relating sunspot activity to human affairs. As a result he founded the (in Russia) respected science of heliobiology: the science dealing with the influence of solar activity and other cosmic factors on the terrestrial biosphere. This definition is his, not mine. If it sounds vaguely familiar, remember that the practice of astrology was harshly dealt with in the former Soviet Union.

But relating the sunspot cycle to events on Earth is not a new idea. A Soviet scientist, V. Desyatov, related suicides and automobile accidents to solar flares, finding that on the second day after solar flares suicides were up by a factor of four to five and automobile accidents were up by a factor of four ("Inhabited Space" edited by D.P. Konstaninov and V.D. Pekelis in NASA TT F-819, NASA Technical Translation, 1975, pp. 146-148). And, other investigators have shown a statistically significant relationship between the solar cycle and such diverse elements as air temperature, precipitation, crop yields, fish catches, and the morbidity in man from such diseases as malaria, plague, tularemia, whooping cough, and influenza ("Solar Activity and Sudden

Changes in the Natural Processes on Earth, A Statistical Analysis" edited by I.P. Druzhinin and N.V. Khamaova in NASA TT F-652, NASA Technical Translation, 1976). For example, in "Inhabited Space," Yagodinskiy relates mortality from whooping cough to the magnetic pertubations of the Earth's field. This latter effect is itself related to solar activity.

Now no one, least of all the Russian investigators, are claiming any causality. What is desired and what has been found is a strong statistical correlation between celestial phenomena and events on Earth. The regularity of celestial phenomena allow predictions of other phenomena and through this latter phenomena predictions of the mundane events of interest. For example, the sunspot activity occurs in regular eleven year cycles. There is a maxima or a minima every five and a half years. Hence mortality from whooping cough can be predicted by observing sunspot maxima and minima.

But of course our astrologers never observed sunspots, at least not our European or Islamic astrologers, but note that the eleven-year cycle of solar activity corresponds almost exactly to the geocentric period of the planet Jupiter. (This contrasts with Jupiter's sidereal period of almost twelve year.) And a further analysis of the sunspot series isolates not only the eleven-year cycle, but also (among others) a prominent twenty-year cycle. Our astrologers may not have observed the sunspots, but they did observe the planet and noted the mutual aspects between them. These planetary cycles are related directly to solar activity and hence the planetary cycles are related to mundane events on Earth: temperature, precipitation, river runoff, water levels in lakes and rivers and ground water, crop yields, and even diseases in man.

The Islamic astrologer lived too far south to observe the Aurora Borealis. Not so the Chinese. They observed both the Northern Lights and (in a crude manner) sunspot activity. This is important because the Chinese observations confirm a long term trend of minimum solar activity ending early in the seventeenth century. K. Frazier, in the March 6, 1976 edition of Science News, points out that this minimum solar activity coincides almost precisely with the coldest point in the climatic minimum on Earth we now call the Little Ice Age. Another climatic cold period on Earth occurred during the seventh and eighth centuries. Oriental records, extending back some 2,000 years indicate a long term sunspot minimum during this period, also. Indeed, the Chinese records show sunspot gaps from 1403 to 1604 and from 579 to 808!

But we have noted these dates before. A shift of the conjunctions of Jupiter and Saturn into water occurred in 571 and 1425. A shift into fire occurred in 809 and again in 1603. Long term sunspot minimums occur during the water cycle of the Jupiter-Saturn conjunctions. And this cycle coincides with major climatic changes on Earth.

What say you now? The astrological theories mentioned at the beginning of this book no longer seem so absurd. On the contrary, these theories were based in part (at least) on hard scientific facts. A major climatic change, such as a mini ice age, would most certainly imply a momentous event affecting large portions of mankind. The ancient scientist had many erroneous ideas concerning the nature of the world in which he lived. But he cannot be faulted for being a poor observer (at least to the limits of his crude measuring instruments). Astrology is an applied science, combining both observations and the theories of why these observations must necessarily be true. The theories may be in error, but the observations are fact. Man's first attempt to apply mathematics for the purpose of prognosticating the future may have been crude. But they were, in many instances, at least as good as those we get today using highly sophisticated mathematics and large digital computers.

While far from complete, I hope this book has demonstrated the viability and basic scientific nature of classical astrology. Much remains to be done. Separating the wheat from the chaff in astrology will not be an easy task, but it is a necessary one if this noble art is once again to be accepted as a valid intellectual pursuit.

Review Questions

1. On June 13, 1925 were born three male children within a few hours of one another. The first was born at 0652 hours LMT in Alexandria, Virginia (Figure 15); the next was born at 0653 hours LMT in Lisbon, Portugal (Figure 16); and the third was born at 0654 hours LMT in Ishnomaki, Japan (Figure 17). What will be the major differences in the lives of these three individuals? How are these differences indicated in their charts?

2. On October 17, 1986 a total eclipse of the Moon occurred. It began at 16h19.7m UT and ended at 22h16.3m UT. The middle of the eclipse was at 19h18m UT. Assuming each of the men mentioned above is alive and living in his hometown, which of them, if any, will this eclipse affect?

3. How long did the effects of the lunar eclipse of October 17, 1986 last? What were its effects?

4. How should the Sun and Moon be delineated if one or both is the ruler of an eclipse?

Appendix A

The Mathematics of Astrology

Time in Classical Astrology

It is assumed that the readers of this appendix already know how to convert from civil to sidereal time; and to translate civil time in one part of the world to that in another part. Those who lack this knowledge should review. For the rest it will be beneficial to go into a bit of depth concerning the meanings of the terms "hour" and "day".

The word day has two meanings. In the first instance it is a period between two successive transits of the upper meridian by the sun. In the second instance it is the period of time between sunrise and sunset. The first definition is called a tropical day or an equinoctial day; and the latter definition is called a seasonal day or civil day or an ordinary day.

The word hour also has two meanings. On the equator one degree will pass the horizon, or any other fixed point, in four minutes or 1/360 of 24 hours. This is called an equinoctial period or an equinoctial time. An equinoctial hour is 15 equinoctial periods, or 60 minutes. Off the equator, however, and hour will be longer or shorter than 60 minutes depending upon the time of year. Just how much longer or shorter is determined by considering the length of the seasonal day because every seasonal day has exactly 12 seasonal hours. Seasonal hours are also called civil or ordinary hours.

For example, on July 4, 1973 at 40 degrees north latitude (just south of New York City) sunrise occurs at 0436 hours (4:36 a.m.), and sunset at 1932 hours (7:32 p.m.). Hence there are 14m 56s of daylight (using tropical, or equinoctial, hours). But this time interval is 12 seasonal hours; so each seasonal hour of daylight equals 1h 14m 40s tropical hours. As there are 24 tropical hours in a day, there must be 9h 04m of nighttime on July 4, 1963 at this latitude. Or each seasonal hour of nighttime equals 45m 20s tropical hours. From this it is evident that one seasonal hour of daytime plus one seasonal hour of nighttime is equal to two tropical hours.

It was stated that an equinoctial hour is 15 equinoctial periods. So also a seasonal hour is 15 seasonal periods. But it is obvious that the seasonal period is not the same length of time as the equinoctial period. In the above example the seasonal period for daytime is 0.98 minutes longer than the corresponding equinoctial period, and the seasonal period for nighttime is therefore 0.98 minutes shorter. In classical astrology the length of the seasonal period is called a horary magnitude or a horary period. It is measured by the number of equinoctial periods that are required to make one seasonal hour. In the above example 18.66 equinoctial periods make an hour (seasonal) of daytime, and 11.33 equinoctial periods make one hour of nighttime. Therefore it is said that the daytime hours of July 4, 1973 have 18.66 horary periods, and the horary magnitude of nighttime hours on this date is 11.33. The latitude of 40 degrees is implied in both measurements of course.

Let s be the hour angle of the sun at its setting. Then the time from meridian transit to setting is s. If the very, very small changes in the sun's declination in a 12-hour period are neglected, then the length of time from sunrise to sunset is 2s. The value for s can be determined from:

$\cos s = - \tan \phi \tan \delta$

Also let s' be the time required for the sun to get to transit from sunrise. Then:

$\cos s' = - \tan \phi \tan \epsilon \sin \alpha_{\odot R} = \cos s$

Where $\alpha_{\odot R}$ is the right ascension of the sun at the time it is rising, and ϵ is the obliquity of the ecliptic? The obliquity of the ecliptic is the angle between the ecliptic (apparent path of the sun) and the celestial equator. This is almost a constant 23°27'.

As an example let us determine the length of daytime at San Francisco, California on August 14, 1974. From the ephemeris the declination of the sun at this time is 14° N 24'. The latitude of San Francisco is 37° N 47'. Therefore:

s = 101° 29'

= 6h 45m 56s

and the hours of daytime are 2s or 13h 31m 52s.

Again, on January 10 the right ascension of the Sun is 19h 23m 49s. What is the length of daytime at Greenwich, England? Of nighttime? What are the horary magnitudes?

The latitude of Greenwich, England is 51° N 29'. So applying the formula:

s' = 59° 24'

Converting s' from degrees to hours, and multiplying by 2, the number of hours of daylight, d, is:

d = 2s' = 7h 55m 12s

From which the number of hours of night, n, are:

n = 24h - d = 16h 04m 48s

The horary magnitude is 1.25 times the number of hours of daylight or nighttime. Therefore on January 10 at the latitude of Greenwich the number of horary periods of day are 9.9, and of nighttime 20.1.

Interpolation

In elementary mathematics interpolation is the process of computing intermediate values of a function from a set of given tabular values of that function. For example, the process of finding a planet's position for a time other than 1200 E.T. (or 0000 E.T. in the case of an ephemeris computed for midnight) is interpolation. Modern astrologers have therefore been doing interpolation for quite some time. The purpose of including this section is to present more precise methods of interpolation than most astrologers are aware of; and more importantly to introduce the concept of differences which are required for calculating the velocity and acceleration of a planet's motion as observed from the earth.

Let $y_0, y_1, y_2, \ldots$ be the values of a function $y = f(x)$ corresponding to equally spaced values of $x_0, x_1, x_2, \ldots$ of the independent variable x. For example, in a typical astrological ephemeris the independent variable, x, is the day of the year; and the dependent variable, y, is the position of a planet or the sidereal time. The following notation will be used.

$y_1 - y_0 = \Delta y_0$
$y_2 - y_1 = y_1$
$y_n - y_{n-1} = y_n$

For obvious reasons, *Δy is called the first forward differences, or simply the first differences. The second difference is the difference of the differences. Namely*

$\Delta y_1 - \Delta y_0 = \Delta^2 y_0$

$\Delta y_2 - \Delta y_1 = \Delta^2 y_1$

and in general the (n+1) st order difference is the difference of the n th order differences:

$\Delta^n y - \Delta^n y_0 = \Delta^{n+1} y_0$

$\Delta^n y_2 - \Delta^n y_1 = \Delta^{n+1} y_1$

A difference table is set up according to the following scheme:

x	y	Δy	Δ²y	Δ³y	Δ⁴y
x_0	y_0				
		Δy_0			
x_1	y_1		$\Delta^2 y_0$		
		Δy_1		$\Delta^3 y_0$	
x_2	y_2		$\Delta^2 y_1$		$\Delta^4 y_0$
		Δy_2		$\Delta^3 y_1$	
x_3	y_3		$\Delta^2 y_2$		
		Δy_3			
x_4	y_4				

As an example, let us set up a difference table for the position of the Moon at noon E.T. for August 12, 1974 through August 16, 1974. The x's are the dates August 12, 13, etc., and the y's are the positions of the Moon at 1200 E.T. for these dates. First we convert the Moon's location from astrological to scientific notation. The table is as follows; and in this table Δx is equal to one day, or 24-hours, for all x.

	x		y	y	Δ²y	Δ³y	Δ⁴y
x_0	Aug 12	y_0	66° 39′ 22″				
				13° 55′ 32″			
x_1	Aug 13	y_1	80° 34′ 54″		25′ 57″		
				14° 21′ 29″		-02′ 01″	
x_2	Aug 14	y_2	94° 56′ 23″		23′ 56″		-03′ 45″
				14° 45′ 25″		-05′ 46″	
x_3	Aug 15	y_3	109° 41′ 48″		18′ 10″		
				15° 03′ 35″			
x_4	Aug 16	y_4	124° 45′23″				

Now suppose that it is required to find the position of the moon for 0400 E.T., August 13. Typical procedures for astrologers would be to note that 0400 is 16-hours after noon. Therefore as 16/24 = 2/3 we have:

$y = y_0 + (2/3) \Delta y_0$

$= 75\ 56' 23''$

From the ephemeris for August 13, 1974 we note that the true position of the Moon at the time in question is 75° 53′ 29″, or almost 3 minutes of arc difference. Some ephemerides (such as Raphael's) give the Moon's position at midnight in addition to the noon position. In such cases the typical method of interpolation gives better results: 75° 54′ 13″ for the case just considered, but still 44 seconds off the true value.

The reason for this can be seen in Figure A1. In the figure y_0 and y_1 represent values of the function corresponding to x_0 and x_1 respectively. We wish to determine the value of y corresponding to the value of x. It is the general practice in astrology to assume a straight line function between y_0 (x) and y_1 (x). If the function is not a straight line, as in Figure A1, then there will be a difference between the true value of y and the value computed using this "straight line" technique.

When more precision is required one must resort to nonlinear interpolation techniques. The formula

$y = y_0 + (x-x_0) \Delta y / \Delta x$

is a linear interpolation formula for the reasons given above. The simplest non-linear interpolation formula is "Newton's Binomial Interpolation Formula". If $\Delta x = x_{n+1} - x_n$ is a constant for all n, then:

$y = y_0 + \sum_{x21}^{n} \Delta^x y \{sx\}$

$= y_0 + \Delta^y_0 s + \Delta^2 y_0 \{s2\} \ldots + \Delta^n y_0 \{sn\}$

where

$s = (x - x_0) / \Delta^x$

and

$\{sx\} = s(s-1)(s-2) \ldots (s-x+1) / x!$

In the formula the $\{sx\}$ is called the interpolation coefficients. A table of these coefficients for Newton's method is given in Table A1.

Normally n = 2 or n = 3 are all that is required for astrological precision. That is the second or third differences give all the precision that will ever be needed for this kind of work.

As an example we shall find the Moon's position in the last example using the second differences and the non-linear.

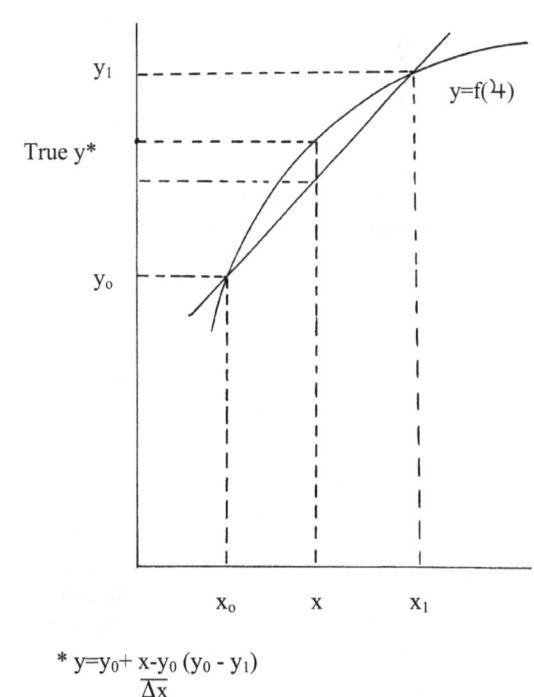

* $y = y_0 + \dfrac{x - y_0}{\Delta x}(y_0 - y_1)$

Figure A1
Non-linear vs Linear Interpolation

Table A1
Interpolation Coefficients (sk) for Newton's Binomial Interpolation Formula

s	(s2)	(s3)	(s4)	(s5)	s
.00	.00000	.00000	.00000	.00000	.00
.01	-.00495	.00328	-.00245	.00196	.01
.02	-.00980	.00647	-.00482	.00384	.02
.03	-.01455	.00955	-.00709	.00563	.03
.04	-.01920	.01254	-.00928	.00735	.04
.05	-.02375	.01544	-.01139	.00899	.05
.06	-.02820	.01824	-.01340	.01056	.06
.07	-.03255	.02094	-.01534	.01206	.07
.08	-.03680	.02355	-.01719	.01348	.08
.09	-.04095	.02607	-.01897	.01483	.09
.10	-.04500	.02850	-.02066	.01612	.10
.11	-.04895	.03084	-.02228	.01733	.11
.12	-.05280	.03309	-.02382	.01849	.12
.13	-.05655	.03525	-.02529	.01958	.13
.14	-.06020	.03732	-.02669	.02060	.14
.15	-.06375	.03931	-.02801	.02157	.15
.16	-.06720	.04122	-.02926	.02247	.16
.17	-.07055	.04304	-.03045	.02332	.17
.18	-.07380	.04477	-.03156	.02412	.18
.19	-.07695	.04643	-.03261	.02485	.19
.20	-.08000	.04800	-.03360	.02554	.20
.21	-.08295	.04949	-.03452	.02617	.21
.22	-.08580	.05091	-.03538	.02675	.22
.23	-.08855	.05224	-.03618	.02728	.23
.24	-.09120	.05350	-.03692	.02776	.24
.25	-.09375	.05469	-.03760	.02820	.25
.26	-.09620	.05580	-.03822	.02859	.26
.27	-.09855	.05683	-.03879	.02893	.27
.28	-.10080	.05779	-.03930	.02924	.28
.29	-.10295	.05868	-303976	.02950	.29
.30	-.10500	.05950	-.04016	.02972	.30
.31	-.310695	.06025	-304052	.02990	.31
.32	-.10880	.06093	-.04082	.03004	.32
.33	-.11055	.06154	-.04108	.03015	.33
.34	-.11220	.06208	-.04129	.03022	.34
.35	-.11375	.06256	-.04145	.03026	.35
.36	-.11520	.06298	-.04156	.03026	.36
.37	-.11655	.06333	-.04164	.03023	.37
.38	-.11780	.06361	-.04167	.03017	.38
.39	-.11895	.06384	-.04165	.03007	.39
.40	-.12000	.06400	-.04160	.02995	.40

s	(s2)	(s3)	(s4)	(s5)	s
.41	-.12095	.06410	-.04151	.02980	.41
.42	-.12180	.06415	-.04138	.02962	.42
.43	-.12255	.06413	-.04121	.02942	.43
.44	-.12320	.06406	-.04100	.02919	.44
.45	-.12375	.06394	-.04076	.02894	.45
.46	-.12420	.06376	-.04049	.02866	.46
.47	-.12455	.06352	-.04018	.02836	.47
.48	-.12480	.06323	-.03984	.02804	.48
.49	-.12495	.06289	-.03946	.02770	.49
.50	-.12500	.06250	-.03906	.02734	.50
.51	-.12495	.06206	-.03863	.02696	.51
.52	-.12480	.06157	-.03817	.02657	.52
.53	-.12455	.06103	-.03769	.02615	.53
.54	-.12420	.06044	-.03717	.02572	.54
.55	-.12375	.05981	-.03664	.02528	.55
.56	-.12320	.05914	-.03607	.02482	.56
.57	-.12255	.05842	-.03549	.02434	.57
.58	-.12180	.05765	-.03488	.02386	.58
.59	-.12095	.05685	-.03425	.02336	.59
.60	-.12000	.05600	-.03360	.02285	.60
.61	-.11895	.05511	-.03293	.02233	.61
.62	-.11780	.05419	-.03224	.02180	.62
.63	-.11655	.05322	-.03154	.02125	.63
.64	-.11520	.05222	-.03081	.02071	.64
.65	-.11375	.05119	-.03007	.02015	.65
.66	-.11220	.05012	-.02932	.01958	.66
.67	-.11055	.04901	-.02855	.01901	.67
.68	-.10880	.04787	-.02777	.01844	.68
.69	-.10695	.04670	-.02697	.01785	.69
.70	-.10500	.04550	-.02616	.01727	.70
.71	-.10295	.04427	-.02534	.01668	.71
.72	-.10080	.04301	-.02451	.01308	.72
.73	-.09855	.04172	-.02368	.01548	.73
.74	-.09620	.04040	-.02283	.01488	.74
.75	-.09375	.03906	-.02197	.01428	.75
.76	-.09120	.03770	-.02111	.01368	.76
.77	-.08855	.03631	-.02024	.01308	.77
.78	-.08580	.03489	-.01937	.01247	.78
.79	-.08295	.03346	-.01848	.01187	.79
.80	-.08000	.03200	-.01760	.01126	.80
.81	-.07695	.03052	-.01671	.01066	.81
.82	-.07380	.02903	-.01582	.01006	.82
.83	-.07055	.02751	-.01493	.00946	.83
.84	-.06720	.02598	-.01403	.00887	.84

s	(s2)	(s3)	(s4)	(s5)	s
.85	-.06375	.02444	-.01314	.00828	.85
.86	-.06020	.02288	-.01224	.00769	.86
.87	-.05655	.02130	-.01134	.00710	.87
.88	-.05280	.01971	-.01045	.00652	.88
.89	-.04895	.01811	-.00955	.00594	.89
.90	-.04500	.01650	-.00866	.00537	.90
.91	-.04095	.01488	-.00777	.00480	.91
.92	-.03680	.01325	-.00689	.00424	.92
.93	-.03255	.01161	-.00601	.00369	.93
.94	-.02820	.00996	-.00513	.00314	.94
.95	-.02375	.00831	-.00426	.00260	.95
.96	-.01920	.00666	-.00339	.00206	.96
.97	-.01455	.00500	-.00254	.00154	.97
.98	-.00980	.00333	-.00168	.00102	.98
.99	-.00495	.00167	-.00084	.00050	.99
1.00	-.00000	.00000	.00000	.00000	1.00

Interpolation technique:

$$y = y_0 + \Delta y_0 \, s + \Delta^2 y_0 \, \{s2\}$$

$$= 66°\,39'\,22" + (13°\,55'\,32")\,(0.67) + (25'\,01")\,(-0.1105)$$

$$= 75°\,33'\,31"$$

or only 2 seconds of arc difference from the true value! Assuming that the precision of the work warrants it, greater accuracy can be obtained by using higher differences.

Velocity and Acceleration of the Planets

Classical astrology places a lot of emphasis on the speed and acceleration of a planet as observed from earth. The time it takes a planet to go from one point to another in the sky is a measure of speed. If the direction (direct or retrograde) of the planet is indicated then this time is also a measure of the planet's velocity. The shorter the time, the faster the velocity. The velocity, v, is:

$$v = x/t = (x_1 - x_0) / (t_1 - t_0)$$

where x_0 is the initial position of the planet and x_1 is its final position; and t_0 is the time when the planet is at x_0 and t_1 is the time when the planet is at x_1.

Now let us look at the difference table for the Moon again. The Δy's are just the amount of degrees that the moon has moved in one day. Δy_0 states that the Moon moved 13° 55' 32" in one day or an average of 34' 49" in one hour. That is the first difference of a planet's position is its velocity. If the velocity is positive, the planet's motion is direct; and if negative, the planet's motion as viewed from the earth is retrograde.

Acceleration is the rate of change of velocity. The second difference of a planet's position is its acceleration, a.

$a = (v_1 - v_0) / (t_1 - t_0)$

If the acceleration is positive the planet's velocity is increasing; and if negative, then the planet is slowing down. In classical astrology if both the velocity and acceleration of a planet is positive then the planet is said to be increasing in its proper motion. Such a planet has a great deal more power than one that is stationary or retrograde.

The Completion of Aspects

Good astrology requires the ability to predict exactly when two planets will be in some aspect degrees apart. To determine this time proceed as follows:

Given that the location of a planet is at x_0, the location, x_1, at some future time, t, can be determine from

$x_1 = (1/2) a t^2 + v t + x_0$

For two planets to be in an aspect, α, to one another one planet must be at x and the other planet at $x + \alpha$. Since the position 1 x is common to both planets:

$(1/2) a_1 t^2 + v_1 t + x_1 = (1/2) a_2 t^2 + v_2 t + x_2 - \alpha$

And since both planets must arrive at their respective places at the same time we can solve for t:

$(1/2) (a_2 - a_1) t + (v_2 - v_1) t + x_2 - x_1 - \alpha = 0$

and so the time at which the aspect will be completed is the solution of the quadratic equation.

$t = [- (v_2 - v_1) + \sqrt{(v_2 - v_1)^2 + 2 (x_1 - x_2 - \alpha) (a_2 - a_1)}] / (a_2 - a_1)$

If $a_2 = a_1$ then the solution is:

$t = (x_2 - x_1 -) / (v_1 - v_2)$

In the above equations, the subscript 1 refers to the initial position and velocity and acceleration of the first planet, and the subscript 2 refers to these values for the second planet.

The Computation of the Elements of a Solar Eclipse

This procedure will be illustrated here using the eclipse of December 24, 1973 that was discussed in the last chapter.

From a map of the eclipse for the location of interest, in this instance Washington D.C., the approximate time of the middle of the eclipse, T_m, and the semi-duration of the eclipse, D, is ob-

Table A2
Factors for Computing Geocentric Coordinates

φ	S		C		φ	S		C	
±0	.993305	1	1.000000	1	±45	.994972	58	1.001678	59
1	.993306	4	1.000001	3	46	.995030	59	1.001737	58
2	.993310	5	1.000004	5	47	.995089	58	1.001795	59
3	.993315	7	1.000009	7	48	.995147	58	1.001854	58
4	.993322	9	1.000016	9	49	.995205	57	1.001912	58
5	.993331	11	1.000025	12	50	.995262	58	1.001970	58
6	.993342	13	1.000037	13	51	.995320	57	1.002028	57
7	.993355	15	1.000050	15	52	.995377	56	1.002085	57
8	.993370	17	1.000065	17	53	.995433	56	1.002142	56
9	.993387	19	1.000082	19	54	.995489	55	1.002198	56
10	.993406	21	1.000101	21	55	.995544	55	1.002254	55
11	.993427	23	1.000122	23	56	.995599	53	1.002309	54
12	.993449	24	1.000145	24	57	.995652	53	1.002363	53
13	.993474	26	1.000169	27	58	.995705	52	1.002416	52
14	.993500	28	1.000196	28	59	.995757	52	1.002468	52
15	.993528	30	1.000224	30	60	.995809	50	1.002520	50
16	.993558	32	1.000254	32	61	.995859	49	1.002570	50
17	.993590	33	1.000286	34	62	.995908	48	1.002620	48
18	.993623	35	1.000320	35	63	.995956	46	1.002668	47
19	.993658	37	1.000355	37	64	.996002	46	1.002715	46
20	.993695	38	1.000392	38	65	.996048	44	1.002761	44
21	.993733	39	1.000430	40	66	.996092	43	1.002805	43
22	.993772	41	1.000470	41	67	.996135	41	1.002848	42

tained. From these values the approximate times of the beginning of the eclipse, T_b, and the end of the eclipse, T_e, are computed.

D = 75 minutes = 1h 15m
T_m = 13h 55m
$T_b = T_m - D$ = 12h 40m
$T_e = T_m + D$ = 15h 10m

These times are all Greenwich Mean Time (GMT). They can be converted to local time when required by noting the longitude, λ, of Washington D.C. and applying the techniques given in all elementary astrology texts.

The above data is still all that is required in many cases. When greater precision is required it is necessary to convert the geodetic latitude of the location in question to geocentric coordinates. Referring to the Table A2, Factors for Computing Geocentric Coordinates, and noting the latitude of Washington D.C. (φ = 38° 53′), we find the geocentric coordinates to be:

Table A2, Continued

23	.993813	43	1.000511	43	68	.996176	40	1.002890	40
24	.993856	44	1.000554	44	69	.996216	38	1.002930	39
25	.993900	45	1.000598	46	70	.996254	37	1.002969	37
26	.993945	46	1.000644	47	71	.996291	36	1.003006	35
27	.993991	48	1.000691	48	72	.996327	33	1.003041	34
28	.994039	49	1.000730	49	73	.996360	32	1.003075	32
29	.994088	50	1.000788	50	74	.996392	30	1.003107	31
30	.994138	51	1.000838	51	75	.996422	29	1.003138	28
31	.994189	51	1.000889	52	76	.996451	26	1.003166	27
32	.994240	53	1.000941	53	77	.996477	25	1.003193	25
33	.994293	54	1.000994	54	78	.996502	23	1.003218	23
34	.994347	54	1.000048	55	79	.996525	21	1.003241	21
35	.994401	55	1.001103	55	80	.996546	19	1.003262	19
36	.994456	56	1.001158	57	81	.996565	17	1.003281	18
37	.994512	56	1.001215	58	82	.996582	15	1.003299	15
38	.994568	57	1.001271	83	83	.996597	13	1.003314	13
39	.994625	57	1.001328	84	84	.996610	12	1.003327	11
40	.994682	58	1.001386	85	85	.996622	9	1.003338	10
41	.994740	57	1.001444	86	86	.996631	7	1.003348	7
42	.994797	59	1.001502	87	87	.996638	5	1.003355	5
43	.994856	58	1.001561	88	88	.996643	3	1.003360	3
44	.994914	58	1.001619	89	89	.996646	1	1.003363	1
45	.994972		1.001678	90		.993306		1.000001	

Geocentric coordinates referred to the adopted spheroid at latitude ϕ:
$\rho \sin \phi' = (S + H) \sin \phi$,
$\rho \cos \phi' = (C + H) \cos \phi$;
H, the altitude above sea-level in units of the equatorial radius of the Earth, is 0.1567850×10^{-6} x altitude in m, or 0.0477882×10^{-6} x altitude in ft.

$\rho \sin \phi' = S \sin \phi = 0.624363$

$\rho \cos \phi' = C \cos \phi = 0.779460$

From Table A3, Table of Besselian Elements of the Eclipse for the approximate time of the eclipse the following values are obtained:

	x	y	sin d	cos d	m(deg)	1
T_b	-1.239150	0.270039	-0.397440	0.917628	10.095750	0.575043
T_m	-0.614266	0.345500	-0.397427	0.917634	28.840040	0.028629
T_e	+0.010610	0.421401	-0.397640	0.917640	47.586130	0.575163

These values are the result of appropriate interpolations. The column marked "1" is the radius of the penumbra for T_b and T_e, and the radius for the umbrta for T_m. The hourly variations (x' and y')

Table A3, Eclipses, 1973
Besselin Elements of the Annular Eclipse of the Sun December 24

E.T.	Intersection of Axis of Shadow with Fundamental Plan		Direction of Axis of Shadow			Radius of Shadow on Fundamental Plane	
	x	y	sin d	cos d	μ	Penumbra	Umbra
12 00	-1.572403	+.229978	-.397447	.917625	0.09831	.574990	+.028501
10	1.489092	.239981	.397446	.917626	2.59767	.575004	.028515
20	1.405779	.249992	.397444	.917627	5.09703	.575018	.028529
30	1.322465	.260012	.397442	.917627	7.59639	.575031	.028542
40	1.239150	.270039	.397440	.917628	10.09575	.575043	.028554
50	1.155833	.280074	.397438	.917629	12.59511	.575055	.028566
13 00	-1.072516	+.290117	-.397437	.917630	15.09447	.575066	+.028577
10	.989199	.300169	.397435	.917630	17.59383	.575077	.028588
20	.905881	.310228	.397433	.917631	20.09319	.575087	.028598
30	.822562	.320295	.397431	.917632	22.59254	.575097	.028608
40	.739244	.330371	.397429	.917633	25.09190	.575106	.028617
50	.655925	.340454	.397427	.917634	27.59126	.575115	.028625
14 00	-.572607	+.350545	-.397426	.197634	30.09062	.575123	+.028633
10	.489288	.360644	.397424	.917635	32.58998	.575130	.028641
20	.405970	.370750	.397422	.917636	35.08934	.575137	.028648
30	.322653	.380865	.397420	.917637	37.58869	.575144	.028654
40	.239336	.390987	.397418	.917638	40.08805	.575149	.028660
50	.156020	.401117	.397416	.917638	42.58741	.575155	.028665
15 00	-.072705	+.411255	-.397414	.917639	45.08677	.575159	+.028670
10	+.010610	.421401	.397412	.917640	47.58613	.575163	.028674
20	.093923	.431555	.397411	.917641	50.08549	.575167	.028677
30	.177235	.441716	.397409	.917642	52.58485	.575170	.028680
40	.260546	.451886	.397407	.917643	55.08421	.575173	.028683
50	.343855	.462063	.397405	.917643	57.58357	.575175	.028685

of x and y will also be needed. As x' = 6 Δy, we have for each of the times of interest:

	x^1	y1
T_b	0.499602	0.060210
T_m	0.499908	0.060546
T_e	0.499878	0.060924

The longitude of Washington D.C., λ, is 77° W 00'. We shall need for each of the times under consideration the value h = μ - λ.

T_b: h = -66.904250 = 293.09575
T_m: h = -48.159060 = 311.84094
T_e: h = -29.413870 = 330.58613

Table A3, Continued

16 00	+.427162	+.472248	-.397403	.917644	60.08293	.575176	+.028686
10	.510467	.482439	.397401	.917645	62.58229	.575177	.028687
20	.593770	.492639	.397399	.917646	65.08165	.575177	.028688
30	.677070	.502845	.397397	.917647	67.58101	.575177	.028687
40	.760367	.513059	.397395	.917648	70.08036	.575176	.028687
50	.843662	.523281	.397393	.917648	72.57972	.575175	.028685
17 00	+.926954	+.533510	-.397391	.917649	75.07908	.575173	+.028683
10	1.010242	.543747	.397389	.917650	77.57844	.575171	.028681
20	1.093527	.553992	.397387	.917651	80.07780	.575168	.028678
30	1.176809	.564244	.397385	.917652	82.57716	.575164	.028675
40	1.260087	.574504	.397383	.917653	85.07652	.575160	.02671
50	1.343361	.584771	.397381	.917654	87.57588	.575155	.028666
18 00	+1.426632	+.595046	-.397379	.917654	90.07523	.575150	+.028661
10	1.5509899	.605327	.397377	.917655	92.57459	.575145	.028655

$\tan f_1$ 0.004755
$\tan f_2$ 0.004732
μ' 0.261732 radians per hour
d' +0.000012 radians per hour

We then compute for each of these times:

$\xi = \rho \cos \phi' \sin h$
$\eta = \rho \sin \phi' \cos d - \rho \cos \phi' \sin d \cos h$
$\varsigma = \rho \sin \phi' \sin d + \rho \cos \phi' \cos d \cos h$
$\eta' = \mu' \xi \sin d - \varsigma d$

The values of μ' and d' are obtained from Table A3 and have the values:

μ' = 0.261732 radians/hour
d' = 0.000012 radians/hour

	ξ	η	ς	ζ	η
T_b	-0.716987	0.694453	0.032425	0.080027	0.074583
T_m	-0.580697	0.779579	0.228986	0.136088	0.060401
T_e	-0.382804	0.842777	0.374933	0.177712	0.039813

Next for each of the approximate times of the eclipse compute:
$u = x - \zeta$
$v = y - \eta$
$u^1 = x^1 - \zeta$
$v^1 = y^1 - \eta'$
$n^2 = u'^2 + v'^2$, $n > 0$

	u	v	u′	v′	n
T_b	-0.522163	-0.424414	+0.419575	-0.014373	+0.176250
T_m	-0.033569	-0.434079	+0.363820	+0.000145	+0.132365
T_e	+0.393424	-0.421376	+0.322166	+0.021111	+0.104237

And for these same times

$L = 1 - \varsigma \tan f$
$\Delta = (1/n)(uv' - u'v)$
$D = uu' + vv'$
$\sin \Psi = \Delta/L$

where f is taken from Table A3:

$\tan f = \tan f_1 = 0.004755$ for T_b and T_e
$\qquad = \tan f_2 = 0.004732$ for T_m

	L	Δ	D	sin Ψ
T_b	0.574889	0.4420242	-0.212968	0.768917
T_m	0.027545	0.434066	-0.012276	-
T_e	0.573389	0.446198	+0.117848	0.778177

The true middle of the eclipse can now be computed:

$\tau_m = -D/n^2 = 0.092744$ hours
$\qquad = 05m\ 33s.9$
$T'_m = T_m + \gamma_m = 14h\ 00m\ 33s.9$

Note that had the value of 14h 00m been used for T_m instead of the 13h 55m the corrected T'_m would be 14h 00m 35s. This indicates that the true value of the middle of the eclipse is relatively insensitive to the approximate time taken from the eclipse map.

To get the corrected times for the beginning and end of the eclipse first compute $\cos \Psi$ for each of these times from the $\sin \Psi$ computed previously. Then:

$\tau_b = (-L/n) \cos \Psi - D/n$
$\qquad = 19^n\ 58^s.5$
$T'_b = T_b + \tau_b = 12h\ 59m\ 58s.5$
and
$\tau_e = (L/n) \cos \Psi - D/n$
$\qquad = -00m\ 54s.7$
$T'_e = T_e + T_e = 15h\ 09m\ 05s.3$

The degree of obscuration (Figure A3) is determined from data for the middle of the eclipse. Therefore using the corrected value of this time list the Besselian Elements for the middle of the eclipse.

x	y	sin d	cos d	μ	l_1	l_2
-.572607	.350545	-.397426	.917634	30.09062	.575123.	028633

The values of ζ, and η, u, and v are then computed using the formulas given previously, but with the new values of the Besselian Elements:

ζ	η	u	v
-0.569219	0.784563	-0.003388	-0.434018

$m^2 = u^2 + v^2 = 0.188383$
$s = (l_1 - l_2)/(l_1 + l_2) = 0.905150$

The angles A, B, and C are computed from:

$\cos C = (l_1^2 + l_2^2 - 2m^2) / (l_1^2 + l_2^2)$
$\cos B = (l_1 l_2 + m^2) / m(l_1 + l_2)$
$A = 180° - (B + C)$

Finally:

$S = sA + B - s \sin C$

where the angles are measured in radians. And the degree of obscuration, S', is then

$S' = S/\pi = 0.147924$

From these computations it is seen that if the approximate values of T_b, T_m, and T_e from the map of the eclipse are used the error would be only 05m 34s for the middle of the eclipse, and 10m 27s (14%) for the semi-duration. This error in duration translates to only five-months in the effects of the eclipse. The error in the degree of obscuration using the approximate values is nil. Hence it is seen that these extensive calculations are required only when extreme precision is desired.

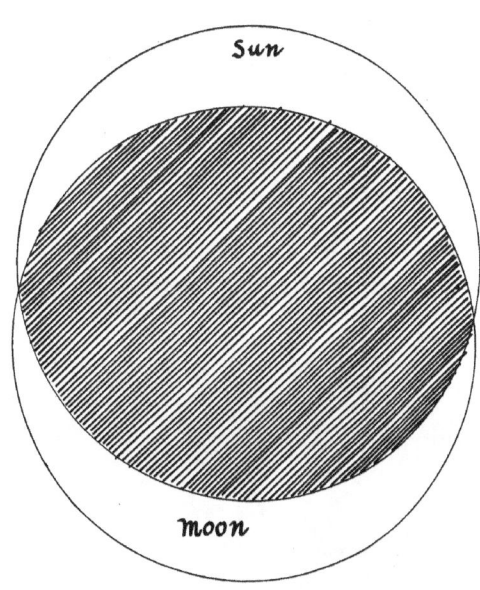

Figure A3. Degree of Obscuration

Appendix B

Rulership of a Point in a Chart

In classical astrology the rulership of a point in a chart such as the Midheaven or a loci (house) cusp considers many more factors than merely the "rulership" of the astrological sign in which the point is posited. Five factors or familiarities are considered: house (domicile), exaltation, term, trine, and decan. These are weighted as follows:

house	= 5
exaltation	= 4
term	= 3
trine	= 2
decan	= 1

The house, or domicile, is merely (in modern astrology) the ruler of the sign in which the point is posited, as is shown in Figure B1. The exaltations are also the same in both modern and classical astrologies: Sun-Aries, Moon-Taurus, Mercury-Virgo, Venus-Pisces, Mars-Capricorn, Jupiter-Cancer, and Saturn-Libra. Tables B1, B2, and B3 give the triplicities (trines), planetary rulers of the terms, and decans. Note that in none of these factors are the trans-Saturian planets of Uranus, Neptune, and Pluto mentioned. In classical astrology these planets cannot rule a point in the chart. In all other respects the procedures of modern astrology apply. For example Uranus is still strongly placed when in the sign of Aquarius.

One more important point. In judicial astrology much work is done with the charts of political entities and their rulers. In such instances the scheme shown above is modified so that the exaltation is given a weight of five and house that of four. Hence it is possible for one planet to rule the Midheaven of the chart of a native when considered as the ruler of a nation, and a different planet

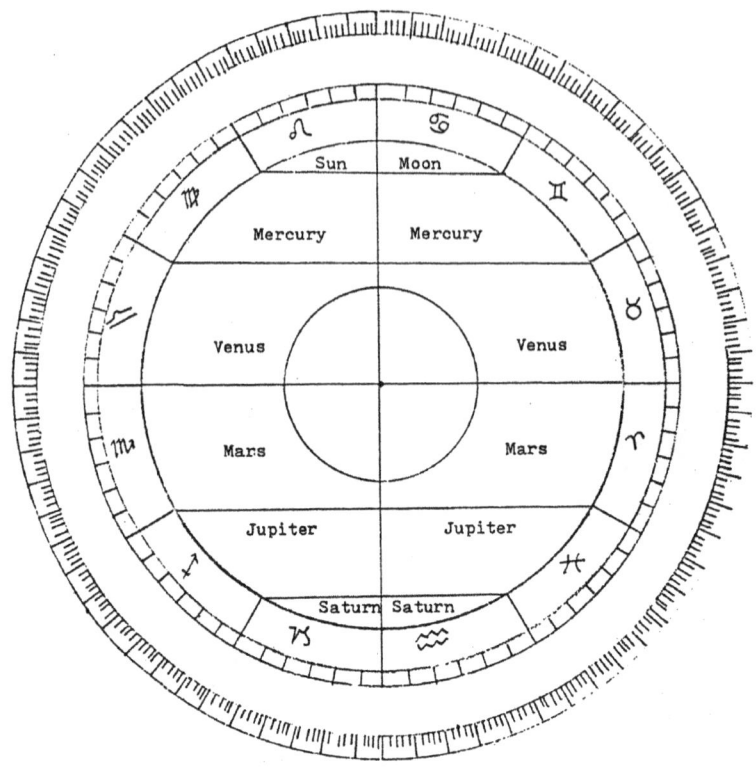

Figure B1. The Houses of the Planets

to rule that same point of that same chart when the native is considered as an individual. For example, let the Midheaven be at 15 Libra in a daytime chart. The Midheaven is then in the house of Venus, the exaltation of Saturn, the term of Venus, and the trine and decan of Saturn. This chart when considered that of an individual has Venus ruling the Midheaven; but when considered as the chart of the ruler of a nation Saturn is the ruler of Midheaven.

Table B1. The Lords of the Triplicities						
		Ptolemy			*Arabic*	
Triplicity	Day	Night	Common	Day	Night	Common
Fiery	☉	♃		☉	♃	♄
Earthy	♀	☽		♀	☽	♂
Airy	♄	☿		♄	☿	♃
Watery	♀	☽	♂	♀	♂	☽

Table B2. The Governors of the Terms

Sign	Term 1		Term 2		Term 3		Term 4		Term 5	
Aries	♃	6	♀	8	☿	7	♂	5	♄	4
	♃	6	♀	14	☿	21	♂	26	♄	30
Taurus	♀	8	☿	7	♃	7	♄	2	♂	6
	♀	8	☿	15	♃	22	♄	24	♂	30
Gemini	☿	7	♃	6	♀	7	♂	6	♄	4
	☿	7	♃	13	♀	20	♂	26	♄	30
Cancer	♂	6	♃	7	☿	7	♀	7	♄	3
	♂	6	♃	13	☿	20	♀	27	♄	30
Leo	♃	6	☿	7	♄	6	♀	6	♂	5
	♃	6	☿	13	♄	19	♀	25	♂	30
Virgo	☿	7	♀	6	♃	5	♄	6	♂	6
	☿	7	♀	13	♃	18	♄	24	♂	30
Libra	♄	6	♀	5	☿	5	♃	8	♂	6
	♄	6	♀	11	☿	16	♃	24	♂	30
Scorpio	♂	6	♀	7	♃	8	☿	6	♄	3
	♂	6	♀	13	♃	21	☿	27	♄	30
Sagittarius	♃	8	♀	6	☿	5	♄	6	♂	5
	♃	8	♀	14	☿	19	♄	25	♂	30
Capricorn	♀	6	☿	6	♃	7	♄	6	♂	5
	♀	6	☿	12	♃	20	♄	25	♂	30
Aquarius	♄	6	☿	6	♀	8	♃	5	♂	5
	♄	6	☿	12	♀	20	♃	25	♂	30
Pisces	♀	8	♃	6	♀	6	♂	5	♄	5
	♀	8	♃	14	☿	20	♂	25	♄	30

Table B3. Lords of the Decans

Sign	Decan 0-10	Decan 10-20	Decan 20-30
♈	♂	☉	♀
♉	☿	☽	♄
♊	♃	♂	☉
♋	♀	☿	☽
♌	♄	♃	♂
♍	☉	♀	☿
♎	☽	♄	♃
♏	♂	☉	♀
♐	☿	☽	♄
♑	♃	♂	☉
♒	♀	☿	☽
♓	♄	♃	♂

Appendix C

The Stars You Are Born Under

In ancient times the stars rising, culminating, and setting at the time and individual was born was considered to be at least as important as the chart or horoscope. These things are the stars one is born under; and they presage a general outline of what an individual's life may be like. Each of the stars has a nature of one of the planets. The effect of the stars you are born under, therefore, is as if those planets were directly on the angles of your chart.

For example Arturus (α-Boo) has the nature of Mars. When rising it portends danger and excitement; when culminating at the birth of an individual it brings riches and honor from military endeavors; and when setting you can look for trouble through rash actions. These are the traditional interpretations given by the ancients. A modern astrologer would say that Arturus presages a varied and interesting life. If on the Midheaven the native could well earn a living through an occupation others would call exciting. Arturus on the Descendant would indicate a possibility of a bad temper or a propensity to act without thinking.

Another example should set the principle of delineation of a person's birth stars. Gemma (or Alphecca) (α-CrB) is of the nature of Venus. When rising it gives the affectionate and sympathetic nature indicative of that planet. But the star, unlike the planet, emphasizes the active rather than the passive elements. Hence Gemma rising portends an individual fond of pleasure. Culminating the star means that the native may prefer pleasure, or pleasurable pursuits, to all other activity. When setting Gemma portends a propensity for pleasure that can, if not checked, bring about real harm to the native: disease, dishonor, and even imprisonment.

Ancient astrologers such as Ptolemy attempted to list each star and its nature individually. Using such a list it has been determined that the ancient astrological significance of the stars can be put in a one-to-one correspondence with the modern spectral class in which the star falls. It has been found that the more oxygen in the star the more beneficial its significance; and the more metallic the star the more malefic it is. The above table gives this correspondence between the stellar spectra and the star's astrological significance. Note that our sun is a star with a class G-spectra. Only the seven planets of antiquity are given. The stellar spectra corresponding to the astrological natures of the trans-Saturian planets has not yet been investigates.

Spectral Class	Astrological Significance
O	Moon
B	Jupiter
A	Venus
F	Mercury
G	Sun
K	Mars
M	Saturn

A short table of risings, culminations, and settings of named starts brighters than magnitude four is appended. In classical astrology the brighter the star the greater will be its effect. So if more than one star may be used it is best to always take the brighter. To use the table enter with the local sidereal time of birth and the geographic latitude of birth. A star whose right ascension (RA) is the same as the local sidereal time is culminating at that time. Stars with the same sidereal time of rising and setting at the appropriate latitude of birth are rising and setting at this time.

For example, a native born at sidereal time 22h 06m at Chicago, Illinois (latitude 41° N 52′) will have Alnair (α-Gru) culminating since the RA of Alnair is 22h 07m. The setting star is Cor Serpentis (α-Ser) as it sets at sidereal time 22h 05m at latitude 40. The closest star to the rising time is Rana (ç-Eri) which rises at 22h 16m at latitude 40 N.

A reasonable orb for the stars you are born under is 15 minutes. In the previous example Rana rises at 22h 16m at latitude 40° N, and at 22h 30m at latitude 50° N. This difference is 14m. Since the native was born at latitude 41° N 52′ (or about 20% of the distance between latitudes 40° N and 50° N) the rising time of Rana at the latitude of birth will be about 22h 19m. This is within the orb, and so Rana is the rising star.

Alnair has a spectra of B5, and is of the nature of Jupiter. Rana and Cor S Erpentis have a spectra of K0 and are of the nature of Mars. Stellar spectra are rated 0 through 9. K0 means that the star is pure Mars. B5 means that the star is beginning to shade towards Venus. These shades of meaning should be accounted for in the delineation.

The table is organized according to the ancient constellations. Modern practice is to put Fomalhaut (α-PsA) in the constellation of Pisces Australis (Southern Fish). Ancient astrologers/astronomers placed this star in Aquarius where it is found in these tables. This distinction is important when considering the effects of constellations in mundane astrology.

Zodiacal stars (stars within five degrees of the ecliptic) are marked by #. Five stars are underlined. These stars make up the ancient Egyptian constrellation of "X". These stars from a double trine in the sky and are very fortunate indeed.

Only named stars brighter than magnitude four are listed in this table. Many unnamed bright stars exist that are not listed. Also it may be necessary to consider latitudes not listed in this table. The rising sand setting sidereal times for any star for any latitude is easily found. Compute:

$t = \cos^{-1}(-\tan\phi \tan\varsigma)/15$

where ς is the start's declinations and ϕ is the geographic latitude. Then for a right ascension, RA, we have

Rising time = RA - t

Setting time = RA + t

Also if

$90 - \phi < \varsigma$

the star will not rise or set at latitude N. It will culminate however. If

$\phi - 90 > \grave{o}$

the star will not be seen above the horizon at latitude ϕ. That is it will neither rise, nor set, nor culminate. These stars are indicated in the table by an asterisk (*).

Sidereal Times of Culminations, Risings, and Settings

Name/magnitude/Spectra					RA		Dec		Latitude 30 Rise		Set		Latitude 40 Rise		Set		Latitude 50 Rise		Set		
					h	m	°	'	h	m	h	m	h	m	h	m	h	m	h	m	
α-UMi	Polaris	2.1	F8		02	07	89	08	**	**	**	**	**	**	**	**	**	**	**	**	Little
β-UMi	Kochab	2.2	K54		14	51	74	17	**	**	**	**	**	**	**	**	**	**	**	**	Bear
γ-UMi	Pherkad	3.1	A2		15	21	71	56	**	**	**	**	*****		**	**	**	**	**	**	
ε-UMa	Alioth	1.7	A0		12	53	56	06	02	56	22	50	**	**	**	**	**	**	**	**	Big
η-UMa	Alkaid	1.9	B3		13	47	49	27	04	57	22	36	02	32	01	01	**	**	**	**	Bear
ξ-UMa	Alula-Australe	3.9	G0		11	17	31	41	03	53	18	40	03	12	19	21	02	08	20	26	
ν-UMa	Alula-Boreale	3.7	K0		11	17	33	14	03	48	18	46	03	03	19	30	01	52	20	42	
α-CVn	Chara	2.9	A0		12	55	38	28	05	06	20	44	04	08	21	42	02	11	23	40	
α-UMa	Dubhe	1.9	K0		11	02	61	54	**	**	**	**	*****		**	**	**	**	**	**	
ζ-UMa	Megrez	3.4	A2		12	14	57	11	02	00	22	28	*****		**	**	**	**	**	**	
β-UMa	Merak	2.4	A0		11	00	56	31**		**	**	**	**	*****		**	**	**	**	**	**	
ς-UMa	Mizar	2.4	A2		13	23	55	04	03	40	23	06	**	**	**	**	**	**	**	**	
γ-UMa	Phad**	2.5	A0		11	52	53	50	02	23	21	21	**	**	**	**	**	**	**	**	
ι-UMa	Talitha	3.1	A5		08	57	48	08	00	17	17	37	22	19	19	35	**	**	**	**	
μ-UMa	Tania-Australis	3.2	K5		10	21	41	38	02	17	18	25	01	08	19	35	**	**	**	**	
λ-UMa	Tania-Borealis	3.5	A2		10	15	43	04	02	04	18	26	00	48	19	42	**	**	**	**	
δ.-Dra	Alib**	3.2	K0		19	13	67	37	**	**	**	**	*****		**	**	**	**	**	**	Dragon
β-Dra	Alwaid	3.0	G0		17	30	52	19	08	17	02	43	**	**	**	**	**	**	**	**	
γ-Dra	Rastaban	2.4	K5		17	56	51	29	08	50	03	02	**	**	**	**	**	**	**	**	
ζ-Dra	Grumium	3.9	K3		17	53	56	53	07	44	04	02	**	**	**	**	**	**	**	**	
ς-Dra	Nodus I	3.2	B5		17	09	65	45	**	**	**	**	**	**	**	**	**	**	**	**	
α-Dra	Thuban	3.7	A0		14	04	64	30	**	**	**	**	**	**	**	**	**	**	**	**	
α-Cep	Aldermin	2.6	A5		21	18	62	29	**	**	**	**	**	**	**	**	**	**	**	**	Cepheus
ρ-Cep	Alphirk	3.3	B1		21	28	70	27	**	**	**	**	**	**	**	**	**	**	**	**	
γ-Cep	Alrai	3.4	K0		23	38	77	30	**	**	**	**	*****		**	**	**	**	**	**	
α-Boo	Arcturus	0.2	K0		14	15	19	19	07	28	21	02	07	07	21	23	06	36	21	54	Plough-man
γ-Boo	Haris	3.0	F0		14	31	38	25	06	42	22	20	05	44	23	18	03	47	01	15	
ε-Boo	Mirak	2.7	K0		14	44	27	11	07	35	21	53	07	02	22	26	06	13	23	15	
β-Boo	Merez	3.6	G5		15	01	40	29	07	03	22	59	05	58	00	04	**	**	**	**	
η-Boo	Mufid	2.8	G0		13	53	18	31	07	08	20	38	06	48	20	58	06	19	21	27	
α-CrB	Gemma	2.3	A0		15	34	26	48	08	26	22	42	07	54	23	14	07	06	00	02	Northern Crown
β-CrB	Nusakan	3.7	F0		15	27	29	11	08	12	22	42	07	35	23	19	06	41	00	14	
β-Her	Kornephoros	2.8	K0		16	29	21	33	09	36	23	22	09	12	23	46	08	37	00	21	Man Kneel-ing
α-Her	Ras Algeth	3.5	M3		17	14	14	25	10	40	23	48	10	24	00	24	10	03	00	25	
δ-Her	Sarin	3.2	A2		17	14	24	52	10	12	00	16	09	38	00	50	08	53	01	35	
β-Lyr	Sheliak	3.4	B2		18	49	35	20	11	20	02	18	10	35	03	03	09	23	04	15	Lyre
α-Lyr	Vega	0.1	A0		18	36	38	46	10	46	02	26	09	47	03	25	07	43	05	29	

Name/magnitude/Spectra					RA		Dec		Latitude						
									30		40		50		
									Rise	Set	Rise	Set	Rise	Set	
				h	m	o	'	h	m h	m h	m h	m h	m h	m	
γ-Lyr	Sulaphat	3.3	A0	18	58	32	39	11	31	02 25	10 48	03 08	09 39	04 17	
ρ-Cyg	Albireor	3.2	K0	19	30	27	54	12	19	02 41	11 44	03 16	10 51	04 06	Bird
α-Cyg	Deneb	1.3	A2	20	41	45	11	12	19	05 03	10 51	06 31	** **	** **	
ε-Cyg	Gienah	2.6	K0	20	45	33	53	13	14	04 16	12 28	05 02	11 12	06 18	
γ-Cyg	Sador	2.3	F8	20	21	40	11	12	24	04 17	11 20	05 22	** **	** **	
η-Cas	Achird	3.6	F8	00	48	57	41	14	24	11 12	** **	** **	** **	** **	Cassi-
β-Cas	Caph	2.4	F5	00	08	59	01	13	12	11 04	** **	** **	** **	** **	opeia
δ-Cas	Rucha	2.8	A5	01	24	60	06	**	**	** **	** **	** **	** **	** **	
α-Cas	Schedir	2.3	K0	00	39	56	24	14	38	10 40	** **	** **	** **	** **	
ε-Cas	Segin	3.4	B3	01	53	63	33	**	**	** **	*****	** **	** **	** **	
β-Per	Algol	2.3	B3	03	07	40	52	19	07	11 06	18 01	12 13	** **	** **	Persues
o-Per	Atiks	3.9	B1	03	43	32	13	20	18	11 08	19 35	11 51	18 28	12 58	
ς-Per	Menkhib	2.9	B1	03	53	31	49	20	29	11 17	19 48	11 58	18 42	13 04	
η-Per	Miram	3.9	K0	02	49	55	48	16	56	12 42	** **	** **	** **	** **	
α-Per	Mirfak	1.9	F5	03	23	49	47	18	31	12 15	15 51	14 55	** **	** **	
α-Aur	Capella	0.2	G0	05	15	45	58	20	48	13 42	19 14	15 16	** **	** **	Chario-
β-Tau	El Nath	1.8	B8	05	25	28	35	22	12	12 38	21 36	13 14	20 43	14 07	teer
ι-Aur	Hasseleh	2.9	K2	04	55	33	08	21	26	12 24	20 42	13 07	19 31	14 19	
ς-Aur	Hoedus I	3.9	K0	05	01	41	02	21	00	13 02	19 53	14 08	** **	** **	
η-Aur	Hoedus II	3.3	B3	05	05	41	12	21	03	13 06	19 56	14 14	** **	** **	
β-Aur	Menkalinan	2.1	A0	05	58	44	57	21	37	14 19	20 10	14 56	** **	** **	
β-Oph	Kelb-Alrai	2.9	K0	17	42	04	35	11	31	23 53	11 27	23 57	11 20	00 04	Serpen-tarius
α-Oph	Ras-Alhague	2.1	A5	17	35	12	35	11	05	00 05	10 52	00018	10 33	00 37	
η-Oph	Sabik	2.6	A2	17	09	-15	41	11	46	22 32	12 03	22 01	10 39	21 55	
ε-Oph	Yed-Posterior	3.3	K0	16	17	-04	38	10	28	22 06	10 33	22 01	10 39	21 55	
δ-Oph	Yed-Prior	3.0	M0	16	13	-03	38	10	21	22 05	10 25	22 01	10 30	21 56	
α-Ser	Cor-	2.7	K0	15	43	06	30	09	28	21 58	09 21	22 05	09 12	22 14	Serpent of Serpantarius
β-Aql	Alshain	3.9	K0	19	54	06	21	13	39	02 09	13 33	02 15	13 24	02 24	Eagle
α-Aql	Altair	0.9	A5	19	50	08	48	13	29	02 11	13 20	02 20	13 07	20 33	
δ-Aql	Deneb Okab	3.4	K0	19	24	03	04	13	17	01 31	13 14	01 34	13 09	01 39	
γ-Aql	Tarazed	2.8	K2	19	45	10	35	13	20	02 10	13 09	02 21	12 54	02 36	
β-Del	Rotanev	3.7	F5	20	36	14	30	14	02	03 10	13 46	03 26	13 24	03 48	Dolphin
α-Del	Sualacin	3.9	B8	20	39	15	49	14	01	03 17	13 44	03 34	13 20	03 58	
γ-Peg	Algenib	2.9	B2	00	12	15	03	17	36	06 48	17 20	07 04	16 57	07 27	Horse
ε-Peg	Enif	2.5	K0	21	43	-09	46	15	20	04 06	15 10	04 16	14 56	04 30	
α-Peg	Markab	2.6	A0	23	04	15	04	16	28	05 40	16 12	05 56	15 49	06 19	
α-Peg	Matar	5.1	G0	22	42	30	05	15	24	06 00	14 46	06 38	13 47	07 37	
β-Peg	Scheat	2.6	M0	23	03	27	57	15	52	06 14	15 17	06 49	14 26	07 40	
ς-Peg	Homan	3.6	B8	22	40	10	42	16	15	05 05	16 03	05 16	15 47	05 32	

Name/magnitude/Spectra					RA		Dec		Latitude 30				Latitude 40				Latitude 50				
									Rise		Set		Rise		Set		Rise		Set		
					h	m	o	'	h	m	h	m	h	m	h	m	h	m	h	m	
γ-And	Almach	2.3	K0		02	02	42	13	17	56	10	08	16	44	11	20	**	**	**	**	Andro-
α-And	Alpheratz	2.1	A0		00	07	28	57	16	53	97	21	16	16	97	58	15	22	08	52	meda
ρ-And	Mirach	2.4	M0		01	08	35	29	17	31	08	45	16	41	09	35	15	15	11	01	
α-Tri	Melellah	3.6	F5		01	52	29	27	18	36	09	08	17	59	09	45	17	03	10	41	Triangle
α-Ari	Hamal#	2.2	K2		02	06	23	21	19	08	09	04	18	41	09	31	18	02	10	10	Ram
β-Ari	Sheratan	2.7	A5		01	53	20	41	19	03	08	43	18	39	09	07	18	06	09	40	
ε-Tau	Ain#	3.6	K0		04	27	19	08	21	41	11	13	21	19	11	35	20	49	12	05	Bull
η-Tau	Alcyone#	3.0	B5		03	46	24	02	20	46	10	46	20	18	11	14	19	38	11	54	
α-Tau	Aldebaran	1.1	K5		04	34	16	28	21	55	11	13	21	37	11	31	21	11	11	57	
27-Tau	Atlas#	3.8	B8		03	38	23	59	20	38	10	38	20	10	11	06	19	30	11	46	
17-Tau	Electra#	3.8	B5		03	43	24	02	20	43	10	43	20	15	11	11	19	35	11	51	
γ-Gem	Alhena	1.9	A0		06	36	16	25	23	57	13	15	23	39	13	33	23	14	13	58	Twins
α-Gem	Castor	1.6	A0		07	33	31	57	00	09	14	57	23	27	15	39	22	21	16	45	
ε-Gem	Mebsuta	3.2	G5		03	42	25	09	23	39	13	45	23	09	14	15	22	26	14	58	
ς-Gem	Mekbuda#	3.9	G0		07	03	20	37	00	13	13	53	23	49	14	17	23	16	14	50	
β-Gem	Pollux	1.2	K0		07	44	28	05	00	35	14	56	23	58	15	30	23	06	16	22	
η-Gem	Tejat Prior#	3.4	M0		06	13	22	31	23	18	13	08	22	52	13	34	22	15	14	11	
μ-Gem	Tejat Posterior	3.2	M0		06	21	22	32	23	26	13	16	23	00	13	42	22	22	14	20	
δ-Gem	Wasat#	3.5	F0		07	19	22	02	00	25	14	13	20	00	14	38	23	24	15	14	
ς-Leo	Adhafera	3.6	F0		10	15	23	33	03	17	17	13	02	49	17	41	02	10	18	20	Lion
γ-Leo	Algieba	2.6	K0		10	19	19	58	03	31	17	07	03	08	17	30	02	36	18	02	
θ-Leo	Coxa	3.4	A0		11	13	15	34	04	36	17	50	04	18	18	07	03	55	18	31	
ρ-Leo	Denebola	2.2	A2		11	48	14	43	05	13	18	23	04	57	18	39	04	35	19	01	
ε-Leo	Ras Elased Australis	3.1	G0		09	44	23	53	02	44	16	43	02	17	17	11	01	37	17	51	
α-Leo	Regulus#	1.3	B8		10	07	12	05	03	39	16	35	03	26	16	48	03	08	17	06	
o-Leo	Subra#	3.8	F5		09	40	10	00	03	17	16	03	03	06	16	14	02	51	16	29	
δ-Leo	Zosma	2.6	A3		11	13	20	38	04	23	18	03	03	59	18	37	03	26	19	00	
γ-Vir	Arich#	2.9	A0		12	40	-05	24	06	43	18	37	06	44	18	36	06	46	18	34	Virgin
δ-Vir	Auva	3.7	M0		12	55	03	31	06	40	19	02	06	42	19	06	06	37	19	11	
ς-Vir	Heze	3.4	A2		13	33	-00	28	07	34	19	32	07	35	19	31	07	35	19	31	
α-Vir	Spica#	1.2	B2		13	24	-11	02	07	50	18	58	08	01	18	46	08	18	18	30	
ε-Vir	Vindemia-trix	2.9	K0		13	01	11	06	06	35	19	27	06	33	19	39	06	07	19	55	
β-Vir	Zavijah#	3.8	F8		11	49	01	54	05	45	17	53	05	43	17	55	05	40	17	58	
α-Lib	Zubenel-Genubi	2.9	A3		14	49	-15	56	09	27	20	11	09	45	19	53	10	09	19	29	Balance
β-Lib	Zubenes-Chamali	2.7	B8		15	16	-09	18	09	38	20	54	09	48	20	44	10	01	20	31	
β-Sco	Acrab#	2.9	B1		16	04	-19	44	10	52	21	16	11	14	20	54	11	45	20	23	Scorpion
α-Sco	Antares#	1.2	M0		16	28	-26	23	11	35	21	21	12	06	20	50	12	53	20	03	
δ-Sco	Dschubba#	2.5	B0		15	59	-22	33	10	54	21	04	11	21	20	37	11	58	20	00	

Name/magnitude/Spectra				RA		Dec		Latitude						
								30		40		50		
								Rise	Set	Rise	Set	Rise	Set	
		h	m	o	'	h	m	h m	h m	h m	h m	h m	h m	
ζ-Sco	Grafias	3.7	K5	16	53	-42	19	13 00	20 46	14 12	19 35	** **	** **	
ν-Sco	Jabbah#	3.9	G0	17	29	-37	17	13 13	21 45	14 08	20 50	15 50	18 08	
λ-Sco	Shaula	1.7	B2	17	32	-37	05	13 15	21 49	14 09	20 55	15 49	19 15	
ς-Sgr	Ascella	2.7	A2	19	01	-29	55	14 19	23 43	14 56	23 06	15 54	22 08	Archer
ε-Sgr	Kaus Australis	1.9	A0	18	23	-34	24	13 56	22 50	14 43	22 03	16 02	20 44	
λ-Sgr	Kaus Borealis	2.9	K0	18	26	-25	26	13 30	23 22	14 00	22 52	14 44	22 08	
δ-Sgr	Kaus Medius	2.8	K0	18	19	-29	50	13 36	23 02	14 14	22 24	15 11	21 26	
o-Sgr	Manubrium#	3.9	K0	19	03	-21	47	13 56	00 10	14 21	23 45	14 57	23 09	
б-Sgr	Pelagus	2.1	B3	18	54	-26	20	14 00	23 48	14 32	23 16	15 19	22 29	
γ-Sgr	Nushaba	3.8	K0	18	04	-30	26	13 23	22 45	14 02	22 06	15 02	21 06	
β-Cap	Dabih#	3.7	G0	20	20	-14	52	14 55	01 45	15 11	01 29	15 34	01 06	Goat
δ-Cap	Deneb Algedi#	3.1	A5	21	46	-16	14	16 25	03 07	16 43	02 49	17 07	02 25	
α-Cap	Gredi	3.8	G5	20	17	-12	37	14 47	01 47	15 00	01 34	15 19	01 15	
γ-Cap	Nashira#	3.8	F0	21	39	-16	47	16 19	02 59	16 38	02 40	17 03	02 15	
ε-Aqr	Albali	3.8	A0	20	46	-09	35	15 08	02 24	15 19	02 16	15 32	02 00	Water Bearer
α-PsA	Fomalhaut	1.3	A3	22	56	-29	45	18 13	03 39	18 51	02 01	19 48	02 04	
α-Aqr	Sadalmelek	3.2	A0	22	04	-00	27	16 05	04 03	16 06	04 02	16 06	04 02	
β-Aqr	Sadalsud	3.1	G0	21	30	-05	41	15 43	03 17	15 49	03 11	15 57	03 03	
δ-Aqr	Scheat	3.5	A2	22	53	-15	57	17 31	04 15	17 49	03 57	18 13	03 33	
η-Psc	Al Pherg	3.7	G5	02	30	+15	13	18 54	08 06	18 37	08 23	18 14	08 46	Fishes
ς-Cet	Baten Kaitos1	3.9	K0	01	50	-10	27	20 14	07 26	20 26	07 14	20 41	06 59	Sea Monster
β-Cet	Diphda	3.2	K0	00	42	-18	07	19 25	05 58	19 46	05 38	20 14	05 10	
γ-Cet	Kaffaljidhma	3.6	A2	02	42	+03	08	20 35	08 49	20 31	08 53	20 27	08 57	
α-Cet	Menkar	2.8	M0	03	01	+03	59	20 52	09 10	20 48	09 14	20 42	09 20	
o-Cet	Mira	2.1	M5	02	18	-03	05	20 25	08 11	20 28	08 08	20 33	08 03	
ε-Ori	Alnilam	1.7	B0	05	35	-01	13	23 38	11 32	23 39	11 31	23 41	11 29	Orion
ς-Ori	Alnitak	2.0	B0	05	39	-01	57	23 44	11 34	23 46	11 32	23 48	11 30	
γ-Ori	Bellatrix	1.7	B0	05	24	+06	20	23 09	11 39	23 03	11 45	22 54	11 54	
α-Ori	Betelgeuse	0.1	M0	05	54	+07	24	23 37	12 11	23 29	12 19	23 18	12 30	
ι-Ori	Hatsya	2.9	O5	05	34	-05	56	23 48	11 20	23 54	11 14	00 02	11 05	
λ-Ori	Heka	3.7	O5	05	34	+09	55	23 11	11 57	23 00	12 08	22 46	12 22	
δ-Ori	Mintaka	2.5	b0	05	31	-00	19	23 32	11 30	23 32	11 30	23 33	11 29	
β-Ori	Rigel	0.3	b8	05	13	-08	14	23 32	10 54	23 41	10 45	23 53	10 33	
κ-Ori	Saiph	2.2	B0	05	47	-09	41	00 10	11 24	00 20	11 14	00 34	11 00	
π₃-Ori	Tabit	3.3	F8	04	48	+06	55	22 32	11 04	22 25	11 11	22 15	11 21	
θ-Eri	Acamer	3.4	A2	02	57	-40	24	22 55	06 59	23 59	05 55	** **	** **	River
α-Eri	Achernar	0.6	B5	01	37	-57	22	23 54	03 20	** **	** **	** **	** **	
β-Eri	Cursa	2.9	A3	05	07	-05	07	23 19	10 55	23 24	10 50	23 32	10 42	

Name/magnitude/Spectra			RA		Dec		Latitude						
							30		40		50		
							Rise	Set	Rise	Set	Rise	Set	
		h	m	o	'	h m	h m	h m	h m	h m	h m	
δ-Eri	Rana	3.7	K0	03 42	-09 51	22 05	09 20	22 16	09 08	22 30	08 54	
μ₂-Eri	Theemin	3.9	K0	04 34	-30 37	23 54	09 14	00 33	08 35	01 33	07 35	
γ-Eri	Zaurak	3.2	K5	03 57	-13 35	22 29	09 24	22 44	09 10	23 04	08 50	
α-Lep	Arneb	2.7	F0	05 32	-17 50	00 15	10 49	00 35	10 29	01 02	10 02	Hare
β-Lep	Nihel	3.0	G0	05 27	-20 47	00 15	10 33	00 38	10 10	01 12	09 36	
ε-CMa	Adara	1.6	B1	06 57	-28 56	02 12	11 44	02 49	11 07	03 43	10 13	Dog
η-CMa	Aludra	1.6	B5	07 23	-09 15	02 38	12 08	03 15	11 31	04 10	10 36	
ς-CMa	Furud	3.1	B3	06 19	-30 03	01 37	11 01	02 15	10 23	03 13	09 25	
β-CMa	Mirzam	2.0	B1	06 22	-17 57	01 05	11 38	01 25	11 19	01 53	10 51	
α-Col	Phakt	2.7	B5	05 39	-34 05	01 11	10 07	01 57	09 21	03 13	08 04	
α-CMa	Sirius	-1.6	A0	06 44	-16 41	01 48	12 07	01 12	11 43	02 38	11 17	
δ-CMa	Wezer	2.0	F8	07 07	-26 21	02 13	12 01	02 45	11 29	03 32	10 42	
β-CMi	Gomeisa	3.1	B8	07 26	+8 20	01 07	13 45	00 58	13 54	00 46	14 06	Procyon
α-CMi	Procyon	0.5	F5	07 38	+05 17	01 26	13 50	01 20	13 56	01 13	14 03	
λ-Vel	Alsuhail	2.2	K5	09 07	-43 20	04 17	13 57	04 51	12 23	05 41	12 33	Argus
ζ-Pup	Azmidiske	3.5	G0	07 48	-24 48	02 50	12 46	03 19	12 17	04 02	11 34	
λ-Car	Canopus	-0.9	F0	06 23	-52 41	03 40	09 06	** **	** **	** **	** **	
β-Car	Miaplacidus	1.8	A0	09 13	-69 37	** **	** **	** **	** **	** **	** **	
ς-Pup	Naos	2.3	A0	08 03	-39 56	03 59	12 07	05 01	11 05	07 47	08 19	
ι-Car	Tureis	2.2	F0	09 17	-59 11	08 17	10 15	** **	** **	** **	** **	
α-Hya	Cor Hydrae	2.2	K2	09 26	-08 31	03 46	15 06	03 55	14 57	04 07	14 45	Water Snake
α-Crv	Alchita	4.2	F2	12 07	-24 35	07 08	17 05	07 37	16 37	08 19	15 55	Raven
δ-Crv	Algorab	3.1	A0	12 29	-16 23	07 08	17 50	07 26	17 32	07 51	17 07	
β-Crv	Kraz	2.8	G5	12 33	-23 16	07 30	17 36	07 58	17 08	08 36	16 30	
ε-Crv	Minkar	3.2	K0	12 09	-22 29	07 04	17 14	07 30	16 48	08 07	15 11	
α-Cru	Acrux	1.6	B1	12 25	-62 58	** **	** **	** **	** **	** **	** **	Centaur
β-Cen	Agena	0.9	B1	14 02	-60 15	** **	** **	** **	** **	** **	** **	
α-Cen	Toliman	0.1	G0	14 38	-60 44	** **	** **	** **	** **	** **	** **	
α-Lup	Arneb	2.9	B2	14 40	-47 17	11 15	18 05	13 01	16 19	** **	** **	Wild Beast
β-Lup	Nihal	2.8	B2	14 57	-43 02	11 07	18 47	12 23	17 31	** **	** **	
α-Gru	Alnair	2.2	B5	22 07	-47 05	18 41	01 33	20 25	23 49	** **	** **	Southern Fish

www.ingramcontent.com/pod-product-compliance
Lightning Source LLC
Chambersburg PA
CBHW081839170426
43199CB00017B/2789